FOODSHED

The
Blooming Fields
www.thebloomingfields.com

Pim & Mary-Ann van Oeveren

RR #1, Site 12, Box 52
Didsbury, AB T0M 0W0
info@thebloomingfields.com

Phone: 403-335-8264
Fax: 403-335-8264
Cell: 403-559-9280

Garden Center - Nursery - Tea / Lunch Room - Gift Shop
Wedding Venue - U-Pick: Fruit, Flowers & Veggies

Foodshed

AN EDIBLE ALBERTA ALPHABET

dee Hobsbawn-Smith

for the Didsbury Library —

Eat your view!

dee

Castor AB

June 2012.

TouchWood
Editions

TouchWood Editions
touchwoodeditions.com

LIBRARY AND ARCHIVES CANADA CATALOGUING IN PUBLICATION
Hobsbawn-Smith, dee
Foodshed : an edible Alberta alphabet / dee Hobsbawn-Smith.

Includes bibliographical references and index.
Issued also in electronic formats.
ISBN 978-1-927129-15-9

1. Agriculture—Alberta. 2. Sustainable agriculture—Alberta. 3. Food
industry and trade—Alberta. I. Title.

S451.5.A4H63 2012   630.97123   C2011-907276-9

Editor: Judy Schultz
Proofreader: Holland Gidney
Design: Pete Kohut
Cover image: Oleg Iatsun, istockphoto.com
All interior photos by dee Hobsbawn-Smith unless otherwise noted.
Author photo: Dave Margoshes

Excerpt printed on page 249 is from the book *A Hunter's Confession* © 2010
by David Carpenter, published by Greystone Books: an imprint of
D&M Publishers Inc. Reprinted with permission from the publisher.
Excerpts printed on pages 1 and 248 are from the book *The Art
of the Commonplace* © 2002 by Wendell Berry, published by
Counterpoint Press. Reprinted with permission from the publisher.
Excerpt printed on page 255 is from the book *The Great Work: Our Way into
the Future* by Thomas Berry, published by Crown Publishing Group
(Harmony/Bell Tower). Reprinted with permission from the publisher.

| | BRITISH COLUMBIA ARTS COUNCIL | | Canada Council for the Arts | Conseil des Arts du Canada | | Canadian Heritage | Patrimoine canadien |

We gratefully acknowledge the financial support for our publishing activities
from the Government of Canada through the Canada Book Fund, Canada
Council for the Arts, and the province of British Columbia through the
British Columbia Arts Council and the Book Publishing Tax Credit.

MIX
Paper from
responsible sources
FSC
www.fsc.org   FSC® C016973

This book was produced using FSC®-certified, acid-free paper,
processed chlorine free and printed with soya-based inks.

1 2 3 4 5 16 15 14 13 12

For my sons, my farmers and Dave.

Peace Country

1. Bridgeview Gardens, Peace River
2. Harmony's Way Farm, Crooked Creek
3. Kemp Honey, High Prairie
4. First Nature Farms, Goodfare
5. Red Willow Gardens, Beaverlodge
6. Nature's Way Veggie Patch, Peace River
7. Dunvegan Gardens, Dunvegan, Edmonton and Fort McMurray
8. Lakeland Wildrice Ltd., Athabasca
9. Four Creeks Ranch, Silver Valley
10. Valta Bison Farms, Valhalla Centre

North Region

1. Swift Aquaculture, Ponoka
2. Bles-Wold Dairy and Bles-Wold Yogurt, Lacombe
3. Birds & Bees Organic Winery and Meadery (formerly En Santé Organic Winery & Meadery), Brosseau
4. Sunworks Farm, Armena
5. The Cheesiry, Kitscoty
6. Sunshine Organic Farm, Warburg
7. Inspired Market Gardens, Edmonton
8. Lola Canola Honey, Bon Accord
9. Sparrow's Nest Organics, Opal
10. Smoky Valley Goat Cheese, Smoky Lake
11. Greens, Eggs & Ham, Leduc
12. Prairie Gardens Adventure Farm, Bon Accord
13. Tipi Creek Farm, Morinville
14. Linda's Market Gardens Ltd., Smoky Lake
15. Sunrise Farm, Killam
16. Sprout Farms Apple Orchards, Bon Accord

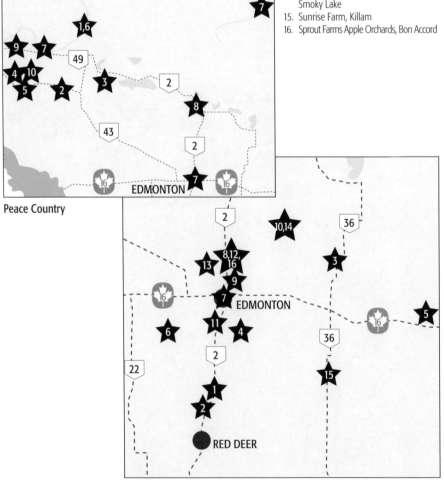

Peace Country

North Region

Central Region

1. Edgar Farms, Innisfail
2. Kayben Farms, Okotoks
3. Field Stone Fruit Wines and Bumbleberry Orchards Inc., Strathmore
4. Country Lane Farms Ltd., Strathmore
5. Winter's Turkeys, Dalemead
6. Elbow Falls Wapiti, Priddis
7. Canadian Rocky Mountain Ranch, DeWinton
8. Greenview Aqua-Farm, Delacour
9. Sun to Earth Farm, Castor
10. Hoven Farms, Eckville
11. TK Ranch, Hanna
12. Chinook Honey Co., Chinook Arch Meadery and Chinook Vinegar Works, Okotoks
13. The Jungle, Innisfail
14. Oxyoke Farms, Linden
15. Gull Valley Greenhouse, Gull Lake
16. Thompson Small Farm and Bergen Farm, Sundre
17. Cakadu Heritage Lamb, Innisfail
18. Sylvan Star Cheese Farm, Red Deer
19. Highwood Crossing Farm Ltd., Aldersyde
20. Eagle Creek Farms Inc., Bowden Sunmaze and Eagle Creek Seed Potatoes, Bowden
21. Poplar Bluff Farm, Strathmore
22. Lund's Organic Farm, Innisfail
23. Hotchkiss Herbs & Produce, Rocky View (s.e. Calgary)
24. Heritage Harvest, Strathmore
25. Hillside Greenhouses, Bowden
26. The Blooming Fields, Didsbury
27. Blue Mountain Biodynamic Farms, Carstairs
28. Buffalo Horn Ranch, Olds

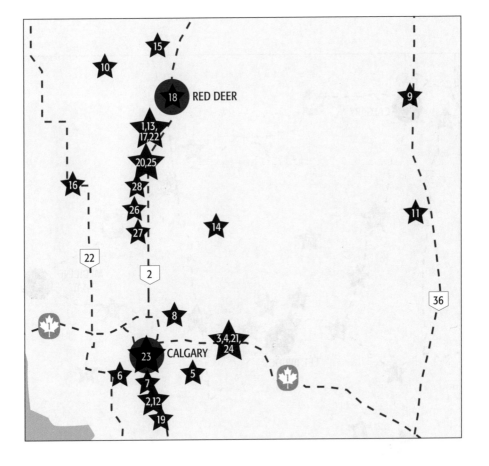

South Region

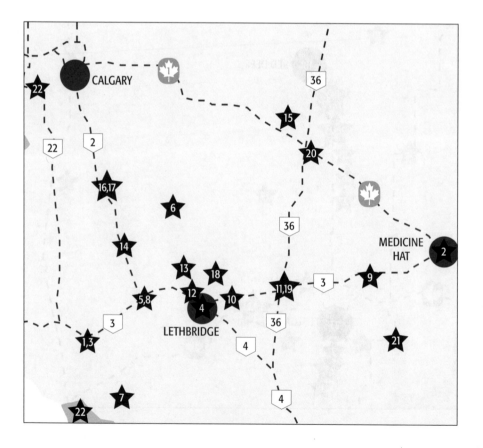

contents

PART ONE
FACES AND FENCES

Author and chef dee Hobsbawn-Smith opened Foodsmith in Calgary's Mission District in 1992, serving locally sourced "Canadiana cuisine." PHOTO: CALGARY HERALD

beginnings

"Eaters must understand that eating takes place inescapably in the world, that it is inescapably an agricultural
act, and that how we eat determines, to a considerable
extent, how the world is used."

—Wendell Berry

The blue flame of my Wolf stove flickered my restaurant, Foodsmith, into being in 1992, in Calgary's historic Mission District, with a new menu every day. A week after I opened the doors, my mother arrived from the family farm west of Saskatoon. She was preceded by my two young sons, who between them lugged an old-fashioned dairy shipper's milk can, one my grandfather had used. "I painted this for you," Mom said. The can was emblazoned with my company logo, a stylized wheat sheaf surrounding the restaurant's name. We positioned it in a place of pride and filled it with the real thing, a sheaf of wheat.

In the early months, a local greenhouse grower knocked on my kitchen door with samples of beautiful organic mesclun strewn with edible flower petals and herbs. Cheese made by Johan Broere arrived by Greyhound bus from Rocky Mountain House. Hutterite farmers brought ducks, the black-hatted men winking at the young women in my kitchen, holding out hand-stitched baby quilts as lures. Then Darrel Winter, a good-natured farmer from Dalemead, sat beside my desk and convinced me to buy the first Winter's Turkey to end up in a restaurant oven.

Our hands were busy with the dozens of daily tasks—chopping herbs, cutting up chickens, making stock, rolling pastry, braising bison—that a professional kitchen must accomplish in order to serve meals to hungry customers. As we worked, one of my sous-chefs, Cape Bretoner John A. McDonald (who took no end of ribbing about his parents' choice of his names!) told stories about his childhood on the East Coast. He said he traded lobster sandwiches for peanut butter—"a step up," he claimed—but I, raised on peanut butter and an Air Force ethos of "waste not," was doubtful. John A. also explained the mysteries of regional maritime specialties like hodgepodge, Solomon Gundy and blueberry grunt.

That first winter, the mesclun grower went bust when a heavy snowfall collapsed the air-supported greenhouse roof. Foodsmith closed in 1994, and I traded my chef's hat for a freelancer's pen, writing about growers, chefs and food.

My connection with my farmer friends deepened. In 1995, I joined the national executive of Cuisine Canada. I remember being dumbfounded when Anita Stewart, the organization's iconoclastic co-founder, said, "Canadian cooks should look beyond lemons and vanilla." Ruling out such simple imported ingredients opened my eyes to what "local" could mean. I began to examine acids in my cooking in a new way, cheering when Alberta's first artisan-made vinegar was finally uncorked in 2010.

My friend Gail Norton owns Calgary's premiere food publication, *City Palate*, and The Cookbook Co. Cooks, an upscale kitchen shop, bookstore and cooking school where I taught for many years. Early in 1998, she tapped my shoulder. "Wouldn't you love to take a busload of city folks out to some of your favourite farms and ranches?" That was the start of the annual Foodie Tootle bus tours. Over twelve years, I took a total of six hundred city folk on all-day trips that culminated in on-farm dinners—locally sourced, of course—to a total of fifty farms and ranches in south-central Alberta. That tour is now in the capable hands of Karen Anderson, who knows a thing or two about buses and hospitality—she owns Calgary Food Tours.

During twelve years under dee's guidance, *City Palate*'s Foodie Tootle took Calgarians to fifty Albertan farms, including Kayben Farms in 2006.

Around the same time, I learned of Slow Food, the international organization that seeks to restore honour and dignity to food and farmers in a fast-paced world. I served on Slow Food Calgary's steering council for ten years. My biggest Slow thrill was chairing the committee that nominated local growers, cooks and culinary students to attend Terra Madre, the Slow Food movement's biennial global get-together in Italy. Many of the farmers profiled in this book have become active in Slow Food and some have taken the mind-blowing trip to Italy. After nominating nearly two hundred Albertans to attend Terra Madre, I attended the conference myself in 2008, and my eldest son, now a Red Seal journeyman cook, went in 2010.

It didn't take going to Italy to know that food is a family matter. Both my sons are confident professional cooks who live in Calgary. I'm the daughter and granddaughter of Saskatchewan dry-land farmers who lived through the Depression and the endless cycles of rising and falling food prices. I sense an internal communion with my late grandmother, Sarah Hofer, who canned, baked bread and cooked every day. A farm field feels like home ground to me. To those who decry a return to that nurturing lifestyle as a step backwards, I can only say that what we have replaced it with is empty.

In 2003, I began to consult about local food for the Alberta government's Dine Alberta program. The goal was to put more Alberta-grown and Alberta-made food on Albertan plates, at home and in restaurants. An annual month-long local-dining festival introduced diners to many local foods, and a publicly accessible database told them whose hands had grown them and where to find them.

Teaching cooking classes and my journalism work, writing for the *Calgary Herald* and Calgary's *City Palate*, led to my fourth book, *Shop Talk*. My previous three cookbooks were published in 1997, 1999 and 2004. *Shop Talk* was a guide to sourcing ingredients in the greater Calgary area, and at its heart was a detailed listing of south-central Albertan growers. When I realized those growers deserved a book of their own, I began travelling around the province to visit my farmers on their home ground, funded in part by a grant from the Alberta Foundation for the Arts.

Foodshed is the culmination of decades of local eating. It celebrates seventy-six Albertan growers. They are ranchers, fishers, farmers, market gardeners and orchardists, forward-looking and forward-thinking. There are many others, equally deserving, who are not included, mostly for reasons of space. Those I have included operate what some might call a throwback to my grandparents' day—mixed farms, market gardens, specialized businesses. Sustainability is a common thread. All are devoted to the important job of feeding the people they know—their neighbours, Albertans.

When I first imagined this A to Z format, it seemed a pretty straightforward approach. It's been more of a sidewinder than I thought, and has led to some imaginative uses of the alphabet. My usage of X, for example, was inspired by my sons' old picture book *Animalia*. I hope you enjoy the slightly goofy thought processes that also landed ducks under Q (for Quackers) and ensured, like any frugal prairie cook, that every letter of the alphabet was used.

My Local

My vision of local food is more inclusive than the one-hundred-mile diet's arbitrary focus. I don't suggest we consume a diet *strictly* based on regionalism, only that we eat *mostly* based on our foodshed. We do live north of the 49th, and our geography imposes limits, so I practise selective, pragmatic self-indulgence instead of self-deprivation. Fair trade coffee, tea, chocolate, olives, vanilla and citrus remain part of my pantry.

It used to be that local was defined by how far a rider on horseback could ride in one day. Market villages were only a few miles apart; produce and meats were consumed within a stone's throw of where they had been raised. In defining local, I imagine concentric rings, starting at my own garden and radiating outwards to include all of Alberta. Where those rings intersect with the ripples of other provinces is how and where waves of change begin. I drink wine, and eat peaches, from BC's Okanagan and Similkameen valleys, a day's drive west of Calgary. In twenty-first-century Canada, with the horsepower harnessed to a car, an Albertan driving from the heart of the prairie can reach, in a day's trip, the Peace Country's boreal forest, the high foothills of Turner Valley, the badlands of the Cypress Hills, the Rocky Mountains, or the northern lakeside arms of the Canadian Shield. That's the Albertan foodshed.

It's not a fad to eat locally, any more than it was in our grandparents' day. It's a simple fact for the majority of the world, as travellers learn. When we go to Toulouse, we eat cassoulet in the style of Toulouse. In Liguria, our pasta is dressed with pesto made from the hillsides' fresh basil, and in Galicia, in northern Spain, we eat anchovies fished from the Bay of Biscay.

In Alberta, a natural foodshed, our pantry overflows with bounty. Albertans raise much of what we need for a pleasurable table: meats, grains, pulses, fruits and vegetables, cheeses, oil and vinegar, even wines made from hardy fruits and berries. Farmers are the wellspring. Better farm practices make for better ingredients. There is a decided correlation: good farming generates good food for good eating, good health and good living.

The new crop of thoughtful growers is tending fields, orchards and water to produce our food in a sustainable manner. Writing in France in 1825 in *The Physiology of Taste*, as translated in 1949 by MFK Fisher, Jean Anthelme Brillat-Savarin said, "Tell me what you eat, and I shall tell you what you are." It is just as true in twenty-first-century Canada. Thanks to travel, television, the Internet and books, we are growing more food-conscious; we are realizing that we are, literally, completely and exclusively, what we eat.

Here's a quick guide to what is produced in this amazing province, Artichokes to Zizania:

A is for . . . artichokes, asparagus, ale, apples.

B is for . . . beans, beets, beef, bison, berries, buffalofish, blueberries, beer, berry wine, butter, broccoli, buffalo mozzarella (*bocconcini di bufala*).

C is for . . . carrots, cheeses, cherries, chard, celery, cantaloupe, casaba, canola, canola oil, chicken, chèvre, cheddar, cucumbers, crabapples, coho, chevon.

D is for . . . dill, dairy, ducks, dried beans.

E is for . . . eggplant, eggs, eels, elk.

F is for . . . fish, fennel, fava beans, feta, field peas, flour, fruit wine, flaxseed and flaxseed oil.

G is for . . . grass-fed beef, garlic, Gouda, goldeye, Gruyère, great northern beans.

H is for . . . horseradish, hemp, honey, honeydew, honeyberries (haskap), habanero chile.

I is for . . . icicle radishes, iceberg lettuce, IPA (India Pale Ale), Italian broccoli, Italian parsley.

J is for . . . juniper, jerky, jacob's cattle beans, jalapeno chile.

K is for . . . kale, kohlrabi, kidney beans, kielbasa.

L is for . . . lettuces, lamb, leeks, lentils, lager.

M is for . . . mustard, milk, mushrooms, mead, muskmelons, mozzarella.

N is for . . . navy beans, nasturtiums, nuts, native grasses, nettles.

O is for . . . onions, oats, oilseeds.

P is for . . . pulses, peas, pumpkin, potatoes, peppers, parsley, parsnips, pickerel, pike, poultry, pears, plums, pierogi, pecorino.

Q is for . . . quail, quark, quackers.

R is for . . . rosemary, raspberries, rapini, rutabaga, rhubarb, radicchio, rabbit.

S is for . . . strawberries, sunflowers, squash, sturgeon, salt, santa claus melon, Santa Fe chile, sausages, savoy cabbage, scotch barley, scotch bonnet chile, serrano chile, smoked fish, snow peas, sorrel, sour cream, soybeans, spaghetti squash, spinach, spelt, spearmint, sugar beets, stinging nettles, stout, swedes, sweet peppers, Swiss chard, salmon, sage.

T is for . . . turnips, tomatoes, turkey, tilapia, trout, Treviso, thyme, tarragon.

U is for . . . u-pick, udon, unsalted butter.

V is for . . . vinegar, veal, vegetables, vegetable oils, verbena, vodka.

W is for . . . watercress, watermelon, whitefish, wine, whisky.

X is for . . . xeriscaping experts, nearly extinct (plains bison).

Y is for . . . yarrow, yard-long beans, yeast, yellow-eyed peas, yogourt.

Z is for . . . zizania, zucchini.

A–Z Recipe List

Food is a circular system. Without consumers to cook and eat their produce, growers are singing to an empty choir. To close the circle, I include twenty-six of my current favourite original recipes. This A to Z collection makes the most of what my farmers, orchardists, fishers and ranchers harvest.

A is for asparagus: Asparagus Roll

B is for berries: Berry Rhubarb Buckle with Yogourt Cream and Berry Ginger Compote

C is for chicken: Butter Chicken

D is for dandelions and other greens: Wilted Greens Agrodolce

E is for elk: Cherry-smoked Elk Loin on Eggplant Salad

F is for fish: Smoked Fish Chowder with Chives

G is for grass-fed beef: Honey-Herb-Cured Alberta Beef or Bison Steak with Spiced Honey Gastrique

H is for honey: Apple-Thyme Mousse and Caramelized Winter Fruit with Filo "Sails"

I is for iceberg and other lettuces: Iceberg, Arugula and Orange Saladio with Pink Pickled Onions and Pine Nuts

J is for jalapeno and other peppers: Desert Stuffed Peppers with Pepitas and Gouda

K is for kale and other sturdy greens: Chicken Ballotine stuffed with Kale, Mushrooms and Sage

L is for lamb: Rogan Josh

M is for milk's immortal leap (cheese): Almost Alsatian Flambée

N is for (great) northern and navy beans and other pulses: Canadian Cassoulet

O is for oilseeds and oil: Herbed Honey Vinaigrette

P is for pork: Charcuterie @ Home: Pork Two Ways

Q is for quackers: Duck Two Ways

R is for roots: Grilled or Roasted Roots and Veggies with Minted Yogourt

S is for squash: Roasted Squash with Peppers, Corn, Feta and Smoked Paprika

T is for tomatoes: Tomato, Walnut and Cilantro Bruschetta

U is for u-pick: Berries in Yogourt Cream with Green Pepper Sauce

V is for vegetables: Vegetable Pakoras with Mint Chutney, Cucumber Raita and Lemon Ketchup

W is for wheat: Flaxseed and Oat Bread Roll with Basil, Gouda and Cold-pressed Canola Oil

X is for xeriscaping experts, nearly extinct (plains bison): Braised Bison Hump with Cherries and Juniper

Y is for yogourt: Yogourt Tiramisu alla Dennice with Fruit Wine or Mead Zabaglione

Z is for zizania: Zizania (Wild Rice) and Cranberry Risotto with Fennel and Sautéed Zucchini

Old-time prairie hay rake and newfangled round bales, the old and the new approach to farming.

My Growers

Much as the great short-story writer Raymond Carver lit up the lives of working-class Americans, my goal is to illuminate the faces and personal lives of my farmers. From the particular, we observe the universal. Times are changing in agriculture, as my friend Kathleen Charpentier, who lives on Sun to Earth Farm near Castor, observes, "Back, but not backwards."

The principle behind *Foodshed* is simple: cook food grown by people you know. As Andreas and Mary Ellen Grueneberg say on their Greens, Eggs & Ham website, "You know your doctor, you know your lawyer, you know your accountant. Who's your farmer?"

Here are my farmers. By monetary standards, a few are wealthy, but most live close to the thin edge of poor. What impels them to work long hours, live rurally and earn a pittance? In a world that professes to admire food and farmers but insists on carping about the high cost of eating, especially local fare, why do these farmers persist in farming? Read on to meet them, and to learn their individual answers.

a is for asparagus

Edgar Farms Innisfail
Doug and Elna Edgar, and Keri and Randy Graham

The tangle of ferns that engulfs my farmyard garden each summer bears no discernible relationship to the sweet green spears that I like to line up on my plate each spring. It is Mother Nature's small joke that we are eating shoots, not leaves, of ferns. Asparagus is hope made tangible, spears spun from fragile ferns and sunshine after winter's absolutist, mineral-fed root vegetables.

On this hot and dusty April day, the wind is whipping sand in taupe gusts across the slope of the hill. "Sand blasting," Elna Edgar calls it, wincing as she pushes her fair hair off her face. One spear, barely three centimetres tall, is emerging from the small, sandy ridge that swells through the Edgars' twenty-eight-acre asparagus field. That solitary spear tells her that it is time to cut last year's old ferns from the field.

Harvest begins in mid-May, and lasts until the end of June when the asparagus stalks bloom into luxuriant ferns. In that brief window of time, Elna, her staff and family will live, breathe, eat and pick asparagus, about twenty thousand pounds worth, over one thousand pounds per acre. In one day in 2006, at the height of the season, they harvested a record amount—nearly two thousand pounds, ten times their initial harvest in 1989. The Edgars had to consider: how big is big enough, and what would they do with any excess crop?

Asparagus was an act of hope. Elna and her husband, Doug, planted their first acre in 1986, despite being told by agriculture experts that asparagus wasn't viable in Alberta's rugged climate, that their crops would be one-fifth of a warmer climate's yield. But that cool weather translates into sweeter vegetables. "It was doing well in the garden. So we put in an acre," Elna explains while driving us to the field. The family waited two years before harvesting to allow the plants to develop. In 1996, they began to add value by pickling the excess crop.

The dream was that the asparagus crop would feed their daughters' future. It did. Cash earned from picking put the Edgars' two girls, Keri and Angie, now in their early thirties, through post-secondary education. "We never believed in just handing money to our kids. We always wanted them to make their own way," Elna recounts. That Keri and her husband, Randy Graham, have returned to the farm with their two young daughters is a triumph too, that their grown children see the value of farm life, and want to raise their girls here. "Keri will decide how big [a farm] is big enough," Elna says.

During harvest, pickers ride through the field—sometimes twice daily—perched three abreast on Doug's ingenious "aspara-buggy," a squat, wheeled device that looks like a distant cousin to a da Vinci flying machine. They sit an eagle's wingspan apart, each positioned over a row. I climb aboard, and am dismayed at how quickly those spears come towards me. I fumble and toss spears untidily into the waiting lug. Beside me, Elna is efficiently snapping and stacking, and steering too, with her feet. When Elna offers me a raw spear, I confess I don't like it raw, but eat it when she insists. Yum. Freshly picked, asparagus tastes like young peas.

In the packing shed, the spears stand in cool water to stay crisp, beside produce drawn from the other family farms that form the successful co-operative known collectively as Innisfail Growers: Beck Farm, The Jungle, Upper Green Farm and Hillside Greenhouses. Shelves and freezers are filled with Elna's pickles, certified Red Angus beef sausages, burgers, jerky, roasts and steaks.

By the time the other fruits and vegetables make their appearance, the Edgars' asparagus season will be done for another year. Her field will return to fern, and rest.

In 2008, Doug and Elna attended Terra Madre, Slow Food's biennial "United Nations of growers and cooks" held in Torino, Italy. The time spent in Italy solidified their sense of connection to the greater community of growers, Elna says.

Before I leave, Elna gives me asparagus crowns as a gift for my garden, welcoming me into that community in a way that merely cooking doesn't impart. Putting the crowns in the ground is the easy part. Patience comes harder.

asparagus roll

This elegant spring dish plays up the sweet earthiness of new-crop asparagus. It takes its seasoning from the classic Italian flavours of *porchetta*—capers, garlic, citrus, parsley, rosemary—using grapefruit's brashness in place of the customary lemon zest. Use pork tenderloin or boneless skin-on chicken thighs. Serves 2.

½ pork tenderloin or 2 chicken thighs, boneless, skin on
1 tsp (5 mL) olive oil or cold-pressed organic canola oil
2 Tbsp (30 mL) minced onion
2 tsp (10 mL) grated ginger root
kosher salt and freshly cracked black pepper to taste
1–2 Tbsp (15–30 mL) herb mustard
½ tsp (2.5 mL) minced fresh rosemary
2 tsp (10 mL) minced fresh parsley
1 Tbsp (15 mL) chopped capers
½ tsp (2.5 mL) grapefruit zest
6 spears asparagus, halved
1 Tbsp (15 mL) chopped oil-cured olives (optional)
1–2 Tbsp (15–30 mL) olive oil or cold-pressed organic
    canola oil, for the pan

Preheat the oven to 400°F (200°C). Cut pork in half and slice each piece of meat into a flat piece about 6 x 6 inches (15 x 15 centimetres), viewing it as a jellyroll to be unwound.

Sauté onion and ginger in oil until tender, about 5 minutes. Season with salt and pepper, then cool. Spread mustard on one side of meat. Sprinkle with herbs, capers, zest and cooked onions. Lay asparagus pieces along one lengthwise edge with chopped olives beside them. Season with salt and pepper, roll up snugly and season the exterior.

Heat a sauté pan and add oil. Place rolls, seam side down, in the hot pan and sear on 4 sides. Transfer the pan to the oven and roast uncovered for 20 to 30 minutes. Remove from the oven, let rest for 5 minutes, then slice in half on a slight angle. Serve with apple yogourt sauce and roasted potatoes.

apple yogourt sauce

olive oil or cold-pressed organic canola oil, for the pan
2 Tbsp (30 mL) minced onion
2 cloves garlic, minced
1 tsp (5 mL) grated ginger root
kosher salt and freshly cracked black pepper to taste
¼ tsp (1 mL) grapefruit zest
juice of 1 red grapefruit
¼ cup (60 mL) dry white wine
½ tsp (2.5 mL) minced fresh rosemary
1 Tbsp (15 mL) minced fresh parsley
¼ cup (60 mL) minced tart apple
1–2 Tbsp (15–30 mL) yogourt

Sauté onion, garlic and ginger in oil. Add salt, pepper, zest, juice and wine. Reduce until the consistency of a thick sauce. Cool. Add herbs, apple and yogourt.

b is for berries

Kayben Farms Okotoks
Claude, Judy, Stephanie, Jolene and Alexis Kolk

In Dijon, *kir*, white Burgundy spritzed with local *cassis*, or blackcurrant liqueur, is the beverage of choice. Judy Kolk's non-alcoholic Albertan cassis—deep purple, rich in vitamin C, antioxidants and anthocyanin—is worthy of its French antecedents. Just add water, juice, soda, champagne or white wine. Of course, a seat in her family farm's café, on its sheltered south-facing patio overlooking the sloping hillside and shrubs, adds to the experience. "My husband, Claude, and I both have the gift of hospitality," Judy says. "We think of people as our guests."

Looking at Judy with her three tall, lean daughters is like looking in a four-way mirror, until they split off to attend to their respective responsibilities. Judy, energetic and professional, leads her staff of young workers with acumen and empathy. Jolene, still a teen, is embarking on a career in pastry arts, steering a course separate from big sisters Stephanie, a chef-in-the-making, and Alexis, a landscape design student. Their diverse interests suit Judy and Claude just fine.

"We wanted to grow a specialty crop on a small acreage, something with overt healthful qualities. But we don't want to give our kids the sense that they are obliged to carry on here. They will have their own lives," Judy says, basking in the late-summer sunlight on the patio.

She and Claude planted three hundred blackcurrant bushes in 2000; they now tend twenty-five thousand shrubs, the heart of a multipronged, twenty-acre empire. A bustling shop filled with local products, Claude's landscaping business and Judy's nursery are fronted by a u-pick berry business and tourist destination that offers school tours and farm animal visits. And a bite to eat. What Judy initially foresaw as a small indulgence people will happily pay for has grown into JoJo's Café, a

Blackcurrants have returned to favour in North America after a forty-year ban for being hosts for white pine blister rust.

bright, airy dining room. It opened in summer 2010, with Stephanie, a SAIT Polytechnic culinary grad, at the helm.

"We all wanted real food, fresh, local. We plan to expand, to include suppers," Judy says, waving at the spacious second floor. It is a big job for a young cook, even with a dedicated bay for the picking in one of the six greenhouses.

What Judy's daughters will make of their own prodigious energy remains to be charted. Stephanie aims to leave Kayben behind in favour of work in Asia or somewhere tropical, but JoJo's will be hard to beat, she says. "My favourite part of the summer is the first hour before anyone else arrives, when it's still cool. I go to the garden with my basket and pick produce for the day."

The orchard includes raspberries and strawberries: berries, like roses, are difficult here, faced with cold winters, hot summers, cool evenings, a short growing season and year-round Chinook winds. Judy plans on planting other hardy fruits—haskap (an Asian berry also called honeyberries, similar to wild blue honeysuckle), cherries and apples. Blackcurrants, the robustly flavoured cousins of the mild gooseberry, are slowly returning to favour in North America after a forty-year ban for being hosts for white pine blister rust. The berries may be helpful in curbing Alzheimer's, according to a New Zealand study published in the *Journal of the Science of Food and Agriculture.* "It appeals, that you can be healthy based on what you eat," Judy says. Their tart, musky, earthy and pungent fruit is as close to being a mineral as a berry can get. Judy has made the most of that forthright flavour. Her store shelves are stacked with jars and bottles of house-made cassis-based products.

A multi-farm, ad hoc organization with several neighbouring businesses, "Flavours of the Foothills," evolved into a self-guided driving tour and cross-marketing tool that saves a little time and work. "People think owning your own business means freedom," Judy says, "but in fact we have little. Someone else can't do my job. My obligation is to my staff and my customers. We've invested too much to walk away. Water, for instance. Six greenhouses need a lot, and we have high-sodium well water, too hard for plants."

The water solution is a moving target; Claude found two tanks that catch runoff from the greenhouses' roofs, and they truck water from Okotoks to supplement the rainwater. The bigger solution will arrive with time. As Judy says with a Gallic shrug, sipping her cassis cooler, "I'm fine. My life's of my own making."

Jolene Kolk works with her mother, Judy, in Kayben Farms' onsite store when she is not immersed in her pastry arts studies.

Elizabeth Chrapko admires a tree planted by her late husband, Vic, in their orchard at Birds & Bees Organic Winery and Meadery near Brosseau.

Birds & Bees Organic Winery and Meadery

(formerly En Santé Organic Winery & Meadery) Brosseau
Tonia, Xina and Elizabeth Chrapko

Conversations with Tonia, Xina and Elizabeth Chrapko are stubbled with "Dad used to—" or "Vic says—". Victor Chrapko was killed in a rural traffic accident in 2008. Tonia and her three siblings grew up on the family farm with an exuberant father whose penchant for experimentation led to the founding of a farmgate fruit winery. "Everything changed after Dad died," Tonia says. "The biggest question was: what will happen to the winery?"

In the 1990s the Chrapkos participated in the University of Saskatchewan's "Fruit for the Prairies" horticulture program, Elizabeth explains during a spring visit to the farm. She halts to sniff a blossoming apple tree. "Vic accepted more and different plants for our growing region—we have thirty-nine cultivars of apples, that's fifteen hundred trees!—and four varieties of sour cherries, four kinds of saskatoons and four varieties of plums." Then there's the pear tree grafted with a multitude of cultivars. Wild blackberries, black cherries, haskap and hazelnuts are the orchard's latest addition. The hazelnuts, slim whippets planted in 2008, are struggling to survive. Even the shelter belt is edible: saskatoons and sea buckthorn.

Vic was a thorn in the Alberta government's side for years, pushing for changes to legislation to allow the existence of cottage wineries. When the law changed in 2005, En Santé became Alberta's first certified organic cottage winery. Vic's inaugural commercial vintage was Calypso Rhubarb, Elizabeth recalls. Mead, slower to ferment, followed. Next was an unusual experiment in *terroir*: when Elizabeth pours a glass for me, I am intrigued by the familiar aroma, but can't identify the beverage as the world's only alfalfa wine, Green Envy. "The Brix [residual sugar content] readings for alfalfa had come back high when we had it analyzed for dairy feed," Tonia says, her brown eyes sparkling behind her glasses. "So Dad said, 'if I can only make wine from what I can grow, I grow alfalfa, therefore . . .'"

Tonia and Xina travelled to British Columbia to meet with other fruit winemakers. They knew that true commitment is born of desire, not obligation. "What our parents instilled in us was a real love for the land," Tonia observes. "The farm was in the family for so long, we didn't want to let it go. We both decided we'd make career changes to keep it for future generations."

Change they did, although Tonia's schedule was diverted by eight months of illness. Xina hit the ground running, and became Alberta's only female winemaker. Her parents' meticulous records had been invaluable when they applied for certified organic status in 1999, and they stood Xina in good stead as she came into the cellar. "It was the school of hard knocks," she says ruefully. "Dad was a purist, he never made fortified wines; and he had decided that Albertan wine should be made by an Albertan. If Dad could do it, we can. We asked the same questions he did. He always said we should give things our best shot and figure them out."

While recovering from her illness in 2008, Tonia, now in charge of marketing, contacted a provincial politician with whom her father had spoken during his crusade for cottage winery status. "I told him there aren't enough hours in the day to have an off-the-farm job as well as making wine—it has to be viable." To be viable, the family wanted their products to be visible, other than just in liquor stores. "People don't go into liquor stores looking for Albertan wines," she says somewhat acerbically. Her timing was perfect: two months later, cottage wineries received permission to sell their products at farmers' markets. In 2011, the family changed the business's name to Birds & Bees Organic Winery and Meadery.

The family has received accolades and awards provincially and internationally, recognizing Victor Chrapko's vision of Alberta. It's ironic, his daughter observes, that as a result of *Endless Feasts*, a PBS film about the family that aired in 2010, more Americans than Canadians know about Victor Chrapko.

Field Stone Fruit Wines and Bumbleberry Orchards Inc. Strathmore
Linden, Marvin and Elaine Gill, and Lorraine and Glen Ellingson

"The days of growing a raw product and making a living are gone. In agriculture, being on the edge means you better make something value-added," Marvin Gill says, looking at the field where several dozen u-pickers are gathering berries or resting in the gazebo. Marvin, his wife Elaine, his brother Linden, and Elaine's sister Lorraine Ellingson and her husband Glen own and operate Alberta's first farmgate fruit winery. Field Stone Fruit Wines sold its inaugural vintage from the farmgate store south of Strathmore in 2005, when provincial laws changed under the advocacy of Victor Chrapko. Three of the partners have off-farm responsibilities as co-owners of Rideau Music in Calgary. However, the partners are all drawing salaries, and have since year three.

"The monetary reward is only one part of it," Marvin says. "We have created a lifestyle that others attach themselves to. People like to be associated with products they know, where they know the farmer." He concedes that the work "has been more intense than we expected. But the net gain is the calibre of our wine and gaining loyal repeat customers."

The first year's five wines, a total of twelve thousand bottles, in four sweet and one off-dry style, were the yield of fifty thousand seedlings planted in 1998. Bumbleberry, a blend of saskatoon, raspberry, strawberry and rhubarb juices, was the opening salvo, and is still the winery's most popular wine. They also produce sweet wines from raspberries, strawberries, saskatoons and wild black cherry (formerly called chokecherry), which has a distinctively spicy endnote.

In subsequent years, the team has added several dry "table" wines, notably cherry, made from Evans, Carmine Jewel and wild black cherries, and oaked blackcurrant. A blackcurrant dessert wine is similar to French cassis.

"After we planted, the hard work started," Marvin recalls. The plan was to obtain organic certification, so they hand-weeded ninety thousand row-feet of saskatoons, and forty-five thousand row-feet of raspberries. "We were in it for the long haul, we waited three to five years for crops, but it was overwhelming, we just about died weeding." They modified their goals, instead following biological ionization principles, an interlocking relationship that dictates strict ratios of minerals to avoid invading insects attracted to stressed plants that are low in sugar.

Marvin pours me a sample of blackcurrant dessert wine in the tasting room, an unpretentious metal building with a stellar view of the orchard and gazebos. He is the farmer and winemaker; fruit wine specialist Dominic Rivard visits to polish and refine the wines. The others share the responsibility for marketing, accounting and attending farmers' markets. "Attending more farmers' markets is our focus," Marvin says. "We can only draw so many people at the farmgate."

Production is currently at fifteen thousand to twenty thousand litres of wine, two-thirds of which is dry or off-dry, and the awards are mounting. The most satisfying win was a 2010 triumph at the Northwest Wine Summit in Oregon, when Field Stone's blackcurrant dessert wine won not only a gold medal but also best in its category.

Marvin hopes to make a *mousseux*, or effervescent, wine, but the timing depends on whether or not the company ramps up volumes to sell wholesale to liquor stores. "If we decide to make larger quantities, maybe we won't do the bubbly. You can only do so much, and only have so much time."

Bridgeview Gardens Peace River
Dan and Gail Marusiak, and Mike and Sheila Marusiak

Dan Marusiak's father, Mike, homesteaded in Rycroft, north of Grande Prairie, in 1927. He was a Ukrainian vegetable peddler, buying produce from other homesteaders and re-selling it. Dan recollects that in the 1940s, when the Alaska Highway was being built, his father sold potatoes to the US government for four cents a pound. "We got four cents a pound in the 1970s, so he did right good," his son says respectfully. Mike had a chance to buy prime flat Peace River bottom land at Dunvegan in the 1940s. Dan remembers, "He was a widower, with a large family, so to look after all the kids, he kept us busy with a big garden. The Yanks, going through to work on the highway, stopped when they got stuck in the mud, and one of 'em took a picture of me, holding a spud as big as my waist."

The Peace River around Dunvegan is a beautiful spot that was already drawing visitors, and Mike made a living as one of two market gardeners in the area. When he died in 1960, Dan inherited the farm and a lot of debts. "After a few years working construction, I was so enthused about gardening. If you could be paid for enthusiasm, I'd have been rich. Whatever they can grow down in Taber, we can grow here," Dan says.

Fruit wines

Fruit wines are a rapidly growing segment of the wine industry, growing in leaps of 25 per cent annually. Marvin Gill believes the growth is driven by several factors: the locavore phenomenon; those who cannot drink grape wine; and wine neophytes who are put off by what Marvin labels "the snob factor" of grape wines.

Fruit wines require no aging. Berries are picked and frozen to accommodate a wide timeframe of ripening and multiple picks. After thawing, the harder berries—saskatoons and chokecherries—are soaked with enzymes in hot water to break down the fruit's fibre and to release their flavours and juices. After filling a barrel press with berry pulp, a water-bladder in the centre pushes the fruit against the slats of the barrel. The juices run out and the remaining pulp is returned to the field as top dressing. From this point, making fortified dessert wine is a brief fermentation to 3 per cent alcohol before vodka is added to elevate the alcohol count to 16 per cent.

He planted five acres of corn and cucumbers, as much as he could tend alone. As the garden grew, he hired neighbour women, then their kids, to hoe and weed and pick. Dan married Gail in 1969, and in the '70s the couple planted strawberries obtained from the Beaverlodge experimental station. "Protem, crossed with wild ones, very hardy and sweet, but they wouldn't travel. We had them for five years, then started getting strawberries from Manitoba. I figured if they grew there, they'd grow here. We had to cover them with straw, but they grew." The Peace Country's first strawberry patch became a big attraction.

Dan and Gail sold the Dunvegan land, which included a historic Hudson's Bay Company factor's house, to the Alberta government in 1984 as a heritage site. For nearly fifteen years, they leased back the land before finding a suitable replacement, more river-bottom land farther east, along the Shaftesbury Trail just outside of Peace River.

Dan remembers that there had been sixteen market gardens along the Trail in the early years, and he had his eye out for one of them, eventually buying two former garden sites along the Peace. He and his son, Mike, spent fifty thousand dollars to put up a game fence to keep a huge herd of marauding deer out of the renewed beds.

Mike the younger, in turn, now tends the garden, now up to fifty acres, with his wife, Sheila. His father drives daily from his Dunvegan home to help. Mike starts eggplant, peppers and cucumbers in a greenhouse for transplantation into an outdoor polytunnel. Outside, in the valley's microclimate, he grows hardy root crops and hot-weather corn, squash, watermelons and cantaloupes.

Dan, now semi-retired, still goes to farmers' markets with Mike, and grows University of Saskatchewan cherries and saskatoons. Market gardening still seems more attractive to him than grain farming. "If it rains, grain farmers can't do anything, they gotta wait for the fields to dry. When it rains and you're a gardener, you put on bigger boots and go to work. What I said a long time ago was, 'man, I'd never do that!' Then I'd say, 'next year I'm gonna get rich.' But you make a living and that's it. You gotta like it or you wouldn't do it."

berry rhubarb buckle with yogourt cream and berry ginger compote

This cross between a cake and a crisp is ideal as a tonic for winter-jaded palates. Change the fruit with the seasons. If you use frozen fruit in the cake, add at least 30 minutes to the baking time. With frozen fruit in the sauce, simmer and thicken it with cornstarch. Serves 12–16.

streusel topping:
¼ cup (60 mL) softened butter
¼ cup (60 mL) brown sugar
¼ tsp (1 mL) ground ginger
½ tsp (2.5 mL) ground cinnamon
⅜ cup (90 mL) flour
½ cup (125 mL) rolled oats

batter:
1 cup (250 mL) all-purpose flour
1 tsp (5 mL) baking soda
¼ tsp (1 mL) kosher salt
½ cup (125 mL) unsalted butter
1 cup (250 mL) sugar
2 large eggs
1 tsp (5 mL) vanilla extract
zest of 1 lemon
2 cups (500 mL) berries, divided
2 cups (500 mL) thinly sliced rhubarb, divided
½ cup (125 mL) minced crystallized ginger
1½ cups (375 mL) yogourt or sour cream

yogourt cream:
3 cups (750 mL) yogourt (choose a gelatin-free brand)
2 Tbsp (30 mL) honey or maple syrup
zest of 1 lemon
2 Tbsp (30 mL) grated fresh ginger root

berry ginger compote:
4 cups (1 L) mixed berries, sliced if large
¼ cup (60 mL) sugar
¼ cup (60 mL) orange juice, mead, fruit wine or late-harvest wine
½ cup (125 mL) minced crystallized ginger
1 tsp (5 mL) grated fresh ginger root

Combine streusel ingredients and mix together by hand. Set aside. Preheat the oven to 375°F (190°C). Lightly butter and flour a 10-inch (22-centimetre) springform pan.

Combine flour, baking soda and salt in a medium bowl. Use a countertop mixer to cream butter and sugar in a large bowl for 1 minute on high speed or until fluffy. Add eggs, vanilla and lemon zest. Mix until combined. Add half the dry ingredients and mix until combined. Add half the yogourt or sour cream and mix until combined. Repeat with remaining dry ingredients and yogourt. Fold half the berries and rhubarb and all the ginger into the batter.

Spread batter into the prepared pan. Top with remaining berries and rhubarb. Crumble streusel evenly overtop. Bake for 60 minutes or until cake springs back when lightly touched in centre. Insert a toothpick in the centre to test for doneness.

Drain the yogourt for 45 minutes through a fine sieve lined with a damp kitchen towel. Discard whey. Add honey or maple syrup, zest and ginger to strained yogourt and mix well.

Mix together compote ingredients and chill, covered, for 1 hour before serving.

To serve, slice cake and garnish with yogourt cream and berry ginger compote.

Hens enjoy a "roost" above the ground.

c is for chicken (and turkey)

Country Lane Farms Ltd. Strathmore
Jerry and Nancy Kamphuis

"I'm not a political guy," Jerry Kamphuis says, "if that means doing what the bulk of people want you to do. But we have been against the grain in what we do for many years. I disagree about antibiotic use and have so since the early 1990s, when I chose not to use antibiotics in the feed for raising our chickens. Industry people say that if you don't use a low level of antibiotics to prevent disease in the poultry, you will end up using more of the stronger antibiotics to cure them. We have not found this to be the case. I also find that weird as I have never taken an antibiotic when I am healthy to prevent catching a disease."

Jerry and his wife, Nancy, have raised birds since 1984 on their farm east of Calgary near Strathmore. They changed their approach to bird production and sales after Nancy urged her nephew to raise a few birds and sell them directly to customers. In 1990, they took their own good advice, and began an Internet-based direct-marketing business that sells their antibiotic-free poultry to nearly five thousand Albertans.

"I left commercial farming because, sitting at the board table of a large corporate processor, I saw its inefficiencies," Jerry recalls. "With direct marketing, I saw an opportunity to reduce prices for our consumers. They pay a lot through the normal retail channel, and I wanted to shorten the steps between producer and consumer."

He's a large, calm man. "I don't need the marketing system," he says frankly, "even though I work within its framework. If the quota system (see p. 27) fell to pieces, I'd still have happy customers coming to me. And they would talk to their MLAs on my behalf—politicians hate controversy."

Jerry says there is no way into the existing system unless you buy existing quota from another farmer. However, the price, should quota become available, is prohibitive, and profit margins for the man at the bottom are marginal at best. "The retailer does the least amount of work and takes the biggest chunk of profit," Jerry says, "while the producer makes pennies."

Every third Monday, Jerry loads his farm truck with live birds, drives from the farm outside Strathmore to the Hutterite plant in Irricana, forty-five kilometres away, where the birds are slaughtered and processed. He returns with a different trailer, and at the farm, Jerry and Nancy cut up some of the birds. On Tuesday, he drives the refrigerated truck and trailer on a nine-pronged delivery route that covers eight hundred kilometres in three days. He stops at church parking lots, malls and hotel lots in designated places in Calgary to meet his customers, and adds Canmore and Red Deer to the route every ten weeks. Everything is pre-ordered online, so Jerry knows what volume to take, plus extra for impulse shoppers.

They started out selling at farmers' markets, but outgrew their table. "A market is a good place to meet people," Jerry says, "but through the Internet, we handle more people and more volume with less effort, and our customers buy five birds at a time. At the markets, people buy one bird this week, and expect you to be there next week so they can buy one more."

Jerry grew up with birds. "I like dealing with livestock. At the other end, I like dealing with the consumer. We have more time freedom, we can travel and family life is better. One of my biggest likes is helping out the Mustard Seed homeless shelter in Calgary. We're able to give them thousands of dollars' worth of chicken per year. Giving back to people in need is something more people should do." It makes a tidy cycle, one that does not rely on quota.

Sunworks Farm Armena
Ron and Sheila Hamilton

The high grass shivers in the breeze, and wild roses bloom along the grass verge where poplars sway and evergreens tower. In the field beyond the yard, calves snooze, piglets socialize and chickens roam. Owners Ron and Sheila Hamilton converge on me with lemonade, and offer to show me Sunworks Farm.

About quota, or chicken by the numbers

In Canada, "quota" is used to stabilize production and markets in dairy, eggs, meat chickens and turkeys. Quota was first issued in 1978 to producers who were in business at the time. According to the Metcalf Foundation's June 2010 paper, the supply management system works well to support established conventional farmers, especially those who produce a single crop. But, as Jerry Kamphuis observes, it does not serve specialists or newcomers.

The quota system is nothing more complicated than a one-time franchise purchase, Jerry says. After speaking with a pair of quota experts, it is a relief to hear a straightforward explanation of the underpinnings of supply management from someone who has seen both sides.

The cost of quota fluctuates with market demand, but the "trackable average" for meat birds is $85 per "unit," says Karen Kirkwood, the general manager of Alberta Chicken Producers. Quota is required to produce more than two thousand birds: smaller producers may only sell their birds at farmgate, at a farmers' market or to an adjacent neighbour. "It's a lot of money upfront," she admits, "but the payoff in the long run is worth it."

Kirkwood attempts to explain a complicated formula that converts the number of "units" into kilograms of chicken, but when she says "multiplied by the percentage of utilization," I bog down. "Each province is allocated production 'units' every eight weeks," she says. "It's all intended to balance the available birds with what is allocated." I roll my eyes. What *is* easily understood is that chicken is big business: in 2011, there were two hundred and eighty registered meat chicken producers in the province of Alberta, each producing more than two thousand birds at a time—a total of over fifty million meat chickens annually. The year before, Albertans consumed seventy-eight pounds (thirty-two kilograms) of chicken per capita.

Sheer volume seems the only way to realize the long-term payoff Kirkwood mentions. A conventional producer is paid a floating rate per kilogram of live weight. (In March 2011, the price was listed on the ACP website as $1.55.) In Alberta, there are only three federally inspected poultry-processing plants for meat shipped out of province. Of seventy-three provincially inspected Albertan poultry slaughterhouses, seventy-one are located in Hutterite colonies. The prices charged by processors and retailers for cleaned chickens are market-driven, but as Jerry Kamphuis observes, retailers make the lion's share of profit with the least amount of effort.

Anyone can keep up to three hundred laying hens without buying or leasing quota. Christina Robinson, Alberta Eggs producers' services manager, says that the cost of laying hen quota ranged in 2010 from $159 to $220, depending on demand. Those unregistered eggs can only be sold at farmgate or at a farmers' market unless the producer pays for the eggs to be graded.

Ron Hamilton admires one of his family's certified organic broiler chickens near a "field barn" on Sunworks Farm near Armena.

Sheila looks tired, but her face is smooth and sunny, a far cry from the thin and chronically ill woman who moved onto the farm in 1992. That year, Ron, an oilfield surveyor with no agricultural experience, moved his wife and daughters, Shae and Erin, from Leduc to a two-hundred-and-forty-acre farm near Armena, southeast of Edmonton. They hoped to grow organic food to aid Sheila's recovery from fibromyalgia and other health issues, and to proactively alleviate Erin's allergies. They immediately began the process of organic certification, and were certified in 1997, which is when the Hamiltons raised their first flock of chicks. The farm is one of only two Albertan certified-organic flocks registered for production in excess of two thousand birds at a time, accounting for less than 1 per cent of Alberta's total production, says Alberta Chicken Producers general manager Karen Kirkwood.

A holistic management course in 1995–96 set the tone. "We are stewards of our children's land," Sheila says. They believe in the ethical and humane treatment of food animals, and raise free-range, grass-fed chickens, pigs, beef cattle and turkeys. In conjunction with Sheila's sister, Dorothy Marshall, a few miles away in Rosalind, they also raise lamb and ducks. Other growers, including Peter and Mary Lundgard of Peace River, raise cattle for the business too.

The chickens and turkeys live in airy movable structures that Ron calls "field barns—bio-secure and predator-friendly," pointing to the electric fence. The buildings sit in the centre of a field that has been grazed down to chicken-friendly height by cattle. Every second day, the structures are towed by tractor to new ground. The birds scratch in fresh grass, roost in the sun and drowse in the shade.

The outdoor areas were originally covered with shade cloth, but metal replaced the fabric to comply with national organic standards implemented after avian flu scares convinced officials that wild birds flying overhead pose a health risk to commercial flocks. In winter, the birds live indoors. "A cold bird is a dead bird," Sheila says plainly.

The system is harmoniously cyclical: the pasture is first covered by the chickens' prolific waste, then grazed by piglets who dig wallows, in the process distributing the chicken litter across the landscape. Moles, digging uphill for fresh air, help even out the moonscape crated by the rooting pigs.

In 2011, Ron and Sheila rethought their overall plan and became what is called a "vertically integrated business." They opened their own certified organic butcher shop and meat-processing plant in Camrose, and put up a slaughterhouse on the farm. "It's what we had to do," Sheila says. "Both are key to our success and longevity. We were relying on outside sources to determine the quality and volume of our birds and meat." She stops and thinks for a moment. "Money used to scare me. It doesn't anymore."

From small eggs, larger beings grow. In 2010, Sunworks raised and marketed one hundred thousand broilers, fifteen hundred turkeys, one hundred and seventy-eight beef cattle, about a million eggs and a smaller number of ducks, geese, hogs, lambs and bison. In 2011, they changed their retail plan as well, leaving the farmers' markets that had been their incubation grounds. Their birds, meat and sausages—celiac and allergy-safe, containing no gluten, wheat, dairy products, nitrates, sulphates or MSG—are retailed at Blush Lane Organics, founded in 2008 by daughter Erin and her husband, Matt Paulsen, with their partners, Rob and Zena Horricks. They're also sold at a pair of independent food stores that cover the east and west sides of Calgary.

"We have always wanted to provide the best food to the most people at the best price possible, and now we can seven days a week," Sheila says. "Everything goes back to holistic management—that's human, social, financial, animal and environmental. We didn't build this business on the premise that our children had to take over. We keep in mind they may not and likely will not. It's never bothered us. Who takes over is not important. That Sunworks carries on is [important]."

Sunshine Organic Farm Warburg
Sherry and Ed Horvath

The conversation about farming is calm and fairly low key until Sunshine Organic Farm co-owner Sherry Horvath starts to talk about the Alberta government's land-appropriation Bills 19 and 36. Then the real kitchen warrior emerges and brandishes her pans. "I'm going to feed those TransAlta representatives our beef, osso buco, when they sit at our table next week to tell us how they are taking our land," the sixtysomething woman says tartly.

At the time, the farm, partway between Leduc and Drayton Valley, was facing land appropriation without compensation under the terms of the new laws. "Lip service in Alberta is paid to different programs, but oil and gas dictate almost everything," she says. "There's a huge coal-fired generating station and mine north of us at Genesee. Farmers are losing their best crop land to it. Sixty quarters were lost in 1979–80, and the next loss is happening as we speak, and they are coming to see us next week. It's taken thirty years to mine what they took over, and our quarter of land will likely be developed ten years from now. But if we settle and sell in 2011, they'll pay a bonus. If not, the bonus is lost."

According to Capitol Power Corporation's Land Use Guidelines (revised November 22, 2010), the current "unit" of the plant is a joint venture between Capitol Power and TransAlta Corporation. "Our intent after purchasing most properties is to decommission the yard site within two years of acquisition. Decommissioning is the process of disposing of surplus items and cleaning up the site in preparation for eventual use by the mine," the guideline reads.

At the same time, Sherry and Ed are facing a suggested high-voltage power transmission line that would parallel the existing line through their land. "I don't spout off at the farmers' market in downtown Edmonton where we sell, I respect other people's opinions, but this needs to be examined."

The land Sherry is intent on preserving is home to pastured Berkshire pigs, grass-finished Black Angus cattle and flocks of pastured chickens and turkeys that the Horvaths sell at the local farmers' market. "It's not what we had expected or planned to do with our lives," Sherry says frankly of the land she and her husband, Ed, have farmed organically since 1981. Ed, a heavy-duty mechanic, had an off-farm job for thirteen years as a gas fitter for a local gas co-op. "Brutal work," Sherry comments, "he was on call 24/7, out until 3:00 AM some nights installing lines. We put all he made into the farm."

Sherry had a successful insurance agency that she sold in 1995. "I love it on the farm, I traded in my briefcase for full-time rubber boots and I wouldn't go back for anything. I like people, and I love that we sell high-quality food to those looking for trust and honesty in their food."

That trust and honesty is lacking in other areas, she thinks. "The rep for the power company came and sat at the table. He was arrogant. It was not a negotiation. The Alberta government wrote in laws that have been passed and put the government above the courts. If we dispute our land being taken, these new bills have taken our right to representation to say this is wrong. The provincial government has given itself the final say. The Charter of Rights always guaranteed our protection, so how can these laws do this? That's dictatorship. I don't think our citizens really recognize how many rights we have lost."

Sherry believes that education effects change. "People need to know their food supply is in jeopardy. Who will feed us if the farms are bought up by oil and gas?"

The whole topic makes any discussion of succession planning for the farm a moot point, I say, but Sherry is more sanguine than I expect. "My feeling is that the door will open when it is supposed to. When it needs to change, it will. I strongly believe that. We have no succession plan at this point, and it's a huge stumbling block. Our daughter Shannon loves the farm; she helps raise the birds. But there is no way she could keep up with the work or buy the farm. If we sell so we can retire, she would lose the land base that grows the seed and crops that feed the animals. There is the heart of the struggle."

In the meantime, the couple has put in a commercial kitchen to produce beef and chicken stock, and ultimately wholesome meals, to go along

Peace Country farmers Larry and Sue King followed the lead of their mentor, organic farmer Jerry Kitt, in raising free-range birds.

with the value-added meats and custom cuts. "We could beat ourselves up every day going, 'oh my gosh, the banker wants a feasibility study and a long-term plan.' In all honesty I see those as a waste of time. Nobody knows what will happen."

With that, Sherry resumes planning the menu for the men who would sit at her table. "When they are taking our land, they gotta know exactly what it is that they are taking away."

Harmony's Way Farm Crooked Creek
Larry and Sue King

Horses have been an integral part of Larry King's life. Now, to his surprise, so are chickens. "We've always had a soft spot for animals," he says wistfully. Larry's spent much of his working life bent over horses' hooves, tacking on shoes, eight hundred horses shod regularly in any one year, he reckons. For twenty-eight years, his day job was on horseback, minding cattle on grazing reserves, from Manyberries south of Medicine Hat to the northern reaches of the Peace Country.

When Larry took a bad fall from a horse he was schooling in 2004, it took months for him to recover, including nearly four months wearing a neck brace. Being unable to work as a farrier for a year was one in a series of catalysts that helped to reshape the family farm.

The farm began with a quarter section of rolling hills in the Clarkson Valley, east of Grande Prairie, that Larry and his wife, Sue, bought in 1986. Sue lived in Fairview with their kids, who began to arrive in 1988, while Larry was away with the cattle, shoeing horses evenings and weekends, and roping whenever he had a chance. In 1995, when the grazing reserves were privatized, Larry came home. They bought the adjoining two hundred acres and moved onto the land. Larry kept on as a farrier to supplement their income.

"It's lucky we ended up with Angus cattle," he says. "They're hardy, good mothers and good eating. We didn't know where we were going at that time; we were still selling at the auction market. We had a dozen sheep too, and sold the lambs conventionally until 2003, and took a loss because there was no accessible organic market."

By 1997, they decided to get serious, investing in a holistic management course in Fairview. In the program, Sue realized that the farm had to work as a whole instead of being unrelated piecemeal projects. A year later, they had the farm certified organic at the suggestion of organic farmer Jerry Kitt, who was mentoring them.

"What I wanted was to have a life," Larry recalls. "To do things I enjoy. We had been working and farming and working and working. What interested me was how to make a living on a small farm, to double the farm by better managing, not by adding more land."

Toward that end, they put up one "hot" electric wire delineating small paddocks. They moved the animals every three days to avoid overgrazing. "Overgrazed land never has the chance to recover," Sue says. "Now we have the lushest pasture you can imagine." They added Berkshire pigs and chickens, pasturing the birds with movable shelters.

The year before Larry took his fall, they began to sell their meats at the Grande Prairie farmers' market. Chickens became the farm's primary source of income, outselling their other meats at the market and at local stores. They added an on-farm processing facility and commercial kitchen, where Sue and Larry custom-cut their own animals and make a dozen

"Turkeys kept us in agriculture," Corinne Dahm says of the heritage breeds she raises with her husband, Darrel Winter.

types of nitrate-free sausages under the watchful eye of master sausage maker Klaus Schurmann (co-owner of the Old Country Sausage Shop in Raymond). A smokehouse went up in 2011, and a farmgate store is next. "An outdoor summer café too," Sue says, "to offer our meats and garden veg from other farmers; simple stuff, not real fancy."

Larry believes that our government should stop importing food that can be raised in Canada. "I think that current regulations favour big business and not small-scale farmers. We communicate by the way we live and by our actions, and we believe that farming is the real enduring economic value when everything else is gone."

They to continue to rely on horsepower. Larry harnesses his Quarter horse/Percheron mares to work the fields or feed the animals, and saddles up whenever the cattle need moving. "Our tractor is old. We aren't mechanical people and we aren't interested in learning," he says wryly. "I don't believe in quads; I believe in horses."

Winter's Turkeys Dalemead
Corinne Dahm and Darrel Winter

Darrrel Winter and his wife, Corinne Dahm, have always seemed like the birds they tend. Darrel, who calls himself "the visiting type," is a benign human version of a sociable "tom" turkey; Corinne is a fine-boned, almost elfin "hen." I have a special fondness for them both: my late restaurant, Foodsmith, was their first restaurant client in 1992. Both talk endearingly of "finding homes" for each of their birds, and sound regretful but deeply appreciative each December when the flocks are gone. "Turkeys kept us in agriculture," Corinne says simply.

Turkey-farming is in Darrel Winter's blood. His dad started a flock in 1958, and sold his birds to meat markets, and privately to consumers. Darrel's brother ran the family business while Darrel studied engineering at the University of Calgary. He and Corinne took over the farm when his brother moved to Saskatchewan with his wife. Their first flock arrived in 1977. Since then, they have scratched out a pioneering organic and free-range turkey run with loyal fans, their birds roosting on restaurant menus and in independent stores.

The business raises two flocks of eleven thousand "poults," or young turkeys, annually. "Value-added is the error of our ways," Darrel says, deadpan. He's only kidding: value-adding has enriched their business since their first sale in my restaurant kitchen. "If you need to sell one thousand birds, you can't just grow one thousand. You need extra. Utility turkeys, and oversupply that don't make it to market, you have to find a home for them." "Home" translates into turkey pastrami, five flavours of smokies and a matching array of smoked salami, plus unsmoked breakfast and cranberry sausages, ground meat and patties in white and dark meat, and breast rendered into "steaks" and "roasts." It's a delicious attempt to teach diners to recognize turkey's value beyond the obvious.

"Darrel's mom told us turkeys are hard to raise. They are fragile, especially as babies," Corinne says, crouched in the centre of her flock. "They take a lot of care. Darrel is up every couple hours in the first few weeks; we watch them 'round the clock."

After attending Slow Food's international cooks' and farmers' gathering, Terra Madre, in 2008 in Torino, Italy, they returned to Alberta intent on making changes on their farm and in their lives. Their first act was planting. "We planted two thousand and ten trees in 2010. It's an act of hope to plant trees." At a picnic table beneath branches that would not reach their full spread for decades, the couple made plans to raise heritage-breed birds.

"Our first attempt was black and white Royal Palms, Cuban natives. We tried incubating some eggs, but the power failed in an early spring storm. We lost all our eggs," Darrel says matter-of-factly. Life and death are closer on a farm than they are in a city. You need to have a backup plan, he says, and he did. "I went to the post office in Strathmore and picked up a box of twenty-five Broad Breasted Bronzes, the same variety Dad started with."

Corinne interrupts. "Little beeping poults! It was like stepping back in history, when all the big items arrived by post!"

The next experiment was fifty Orlopp black turkeys, a breed that's been raised in Ontario for seventy years. "They are friendly and bright, flyers, out of the barn more than in," Corinne says. "They greet you, and talk away to you, and they parade. So calm, so chatty and gregarious, a joy to have, even though they are only with us for a matter of months."

A trip to Cuba in 2010 with Edmonton food activist Ron Berezan was a seminal experience. "We saw community gardens, organic farms, mostly *organipónicos*, reclaimed weed patches and dumps made into food sources." They came back inspired, Corinne says. "It changed how we look at the world and our circle of friends, and what we will do with this land. Because you know something is happening on this planet, and what are you doing? When your grandkids say, 'what did you do about it?' you want to have a good answer." The trip was a benchmark that made them present and accountable for their actions. "There's this song," she continues, "It's called 'Think Like a Mountain.' A mountain thinks in millions of years. For the most part, our society thinks in minutes."

butter chicken

My rendition of this rich Indian classic uses a spice blend (*garam masala* or *ras el hanout*), chicken thighs and fresh tomatoes. Don't flinch at the large amount of cream in this dish, but if you prefer, use plain yogourt instead. (See Rogan Josh, p. 110, for my yogourt-based version of curry.) This is equally delicious made with turkey or pork. Serve with basmati rice, flatbread and lentils. Serves 8–10.

4 Tbsp (60 mL) butter
2 onions, minced
6 garlic cloves, minced, divided
1 Tbsp (15 mL) grated ginger root, divided
4 Tbsp (60 mL) mild garam masala or ras el hanout, divided
8 boneless chicken thighs, cut into 1 in (2.5 cm) cubes
½ cup (125 mL) dry white wine (optional)
4–6 ripe tomatoes, seeded and diced
2 Tbsp (30 mL) tomato paste
1 cup (250 mL) chicken stock
1–2 cups (250–500 mL) whipping cream
lemon juice to taste
kosher salt and pepper to taste
2 Tbsp (30 mL) minced fresh parsley or cilantro

Melt butter in a large, heavy-bottomed pan. Add onions and half the garlic, half the ginger and half the garam masala. Cook gently over medium heat until onions are tender but not browned, about 10 minutes. Stir often. Meanwhile, sprinkle remaining garlic, ginger and garam masala on chicken pieces and mix thoroughly in a bowl.

When onions are tender, add chicken. Stir well and cook for 5 minutes. Add optional wine and cook for 3 to 5 minutes. Add tomatoes, tomato paste, stock and cream. Mix well. Cover and simmer for 1 hour, or until tender, stirring from time to time and adding additional water, stock or cream if liquid level drops too low. Remove the lid and simmer until thickened if sauce is too runny. Taste, adjusting the balance with lemon juice, salt and pepper. Garnish with cilantro or parsley and serve hot.

Dandelions, like chives, make a "dandy" addition to the salad bowl if picked before they flower and go to seed.

d is for dandelions and other greens

Inspired Market Gardens Edmonton
Gwen Simpson

In 1991, when Gwen Simpson toured eastern Europe and Russia, the Berlin Wall had just come down. In Moscow, she saw how little food there was. "Nothing fresh," she says, "peddlers selling vegetables you couldn't identify as vegetables, and people buying them. It was all unrecognizable, rancid or preserved." That trip's indelible images have endured. "Not that scaremongering works," she says vehemently, "but Canadians don't understand what it means to not have good food. And we only spend 11 per cent of our disposable income on food, less than half what western Europeans spend."

Gwen has put a lot of time into thinking about what food—and for how many dollars—Canadians will buy. Her market garden south of Edmonton has been an integral part of her life since 1999. The recent forced sale of the land it occupied has her thinking about costs of land, raw ingredients and transportation of imported food, what people are willing to settle for, and the view consumers have of farmers.

"Everything comes down to numbers. If farmers can't stay in business, they can't stay on the farm. What the public sees is the old story—'oh farmers, they are always moaning, and then they buy a truck.' What they don't see is that old farmers had cheaper land and old machinery. New farmers are heavily in debt or undercapitalized."

After ten years living and gardening in England, Gwen returned as a huge fan of English market gardens, and felt their absence keenly in Edmonton. She and her sister bought twenty-eight acres of rolling hills near Carvel to transform into an organic market garden. Gwen wanted to raise herbs and dried flowers, and visited farms in BC, where Canadian agritourism was already in full flower. In 2004, she made a lovely garden

Salad greens are easily grown in containers of all sizes.

and lovely leaflets. Then she watched people stop in front of her slightly-far-from-the-action yard and not come up the driveway. So she put up a greenhouse and a store right beside the road, and Gwen began to attend farmers' markets.

She learned soon enough that she couldn't sell enough herbs and edible flowers to make a living. She added them to a spicy blend of field greens—mustard, mizuna, dandelion, arugula, Asian choy, purslane, spinach, salad burnet—in a "cut-and-come-again" style, the young plants cut off above the roots and allowed to grow back for subsequent cutting. People bought all that she could grow. She added custom-made soaps, jams and jellies produced using her own herbs, selling them at the farm store and at the markets. But when she finally costed her salad greens, Gwen was shocked to realize that even without adding a wage for herself, she was underselling her salad greens in a big way. "I had been arrogant about farmers not making money. I thought the reason was that farmers didn't know how to market, and they weren't interested in going direct to consumers. Well, I finally realized how small the margins are."

The monetary problem was intensified by complications with the co-owners of the land. Gwen's sister and brother-in-law were increasingly unhappy, and made overtures to a real estate developer, against the express terms of a deal made when the land had been initially subdivided. After two years of argument, they bought Gwen out in 2011, leaving her market garden homeless. "I realized that field greens are extremely difficult to make a living at except in massive quantity. I never wanted to grow three acres of lettuces. I don't know how to acquire another land base. I'm working on it. All I want to do is farm. In my late fifties, I finally figured it out." In 2011, Gwen relocated her business to a greenhouse located on the grounds of the Alberta Hospital. She continues to evaluate her goals—and her limitations.

The larger issue is clear to Gwen: "How do we get the public to be willing to pay more for food? They are so used to cheap, subsidized food; they don't know the real cost of growing it. That's the unanswerable question."

Mesclun is comprised of mixed herbs, greens and edible flowers in a blend that suits the grower.

P is for dandelions and other greens

wilted greens agrodolce

My favourite vegetable dish uses whatever greens I have in the garden or fridge—chard, kale, spinach, mesclun or frisée. Lily gilders could add a dollop of chèvre or cheese sauce, but it is not necessary. Top with poached eggs for a more substantial dish. Serves 2.

½ red, orange or yellow bell pepper
2 slices bacon, in ½ in (1 cm) cubes or strips
½ onion, thinly sliced or minced
1 bunch chard, stalk removed and minced, leaves chopped
2 garlic cloves, sliced
4 oz (120 g) smoked sausage, your choice, diced or sliced
2–3 Tbsp (30–45 mL) dried cranberries
vinaigrette:
2 Tbsp (30 mL) olive oil or cold-pressed canola oil
1–2 Tbsp (15–30 mL) fruit-infused white wine vinegar
1–2 Tbsp (15–30 mL) maple syrup
½ tsp (2.5 mL) minced fresh thyme
kosher salt and freshly cracked pepper to taste
croutons:
2–4 slices baguette
2 Tbsp (30 mL) chèvre

Preheat the broiler. Lay pepper, skin side up, directly beneath the broiler element. When skin is blackened, put pepper in a bowl and cover snugly. Leave in the bowl for 5 minutes, then remove and discard charred peel. Dice the pepper flesh. Cook bacon in a shallow saucepan until almost crisp. Add onion, chard stalks and garlic, setting aside the chard leaves. Add a tablespoon of water to the pan and cook for 5 to 7 minutes, until onion is translucent. Add cranberries and chard leaves or other greens along with 1–2 tablespoons (15–30 millilitres) water. Cover the pan and reduce the heat. Cook for 1 to 2 minutes, or until wilted and tender, depending on the type of greens. Uncover and taste for tenderness. Make vinaigrette while the chard leaves wilt. Add vinaigrette to the pan and mix gently to coat the greens. Spread cheese on the bread, broil until bubbly and hot, and serve with the greens.

e is for elk

Elbow Falls Wapiti Priddis
Win Niebler

Win Niebler brings a city-bred European viewpoint to Albertan ranch life. He raises elk on one hundred and ten acres in a foothills valley between Bragg Creek and Priddis, in what he calls God's Country. "I'm a city slicker from Munich, and now I have a million trees," he says. "I have never regretted it. Alberta is a diamond in the rough, where you can be anything you want to be if you want it hard enough."

Win met his Canadian wife-to-be, Cathy, at Ayers Rock in the early 1980s. "Travel is a good education. I went from ironed shirt and pants on the Munich train to shorts and bare feet in Australia," he says. Three months on the road turned into two years, including time in Southeast Asia. When work in New Zealand proved elusive, Cathy said, "How about Canada? It's like New Zealand, only bigger." They flew to Toronto, picked up her station wagon and drove west. Win, trained in biomedical imaging, worked for ten years for Siemens Electric, and the couple bought the rural residence he now inhabits with his second wife, Pat.

In 1989, when the budding entrepreneur wanted to acquire animals that were native to the landscape, he settled on elk. "Why cross borders into exotic food? The elk were here long before we arrived." Eighteen elk were pricey in those days: seventeen thousand dollars each. Every calf the herd produced, Win sold the next season to get out of debt.

Fencing cost even more than the cows. "They're herd animals. The main issue is to keep wild elk out, so I spent two years putting up six kilometres of eight-foot fence. I did everything myself, with two small kids, from vaccines to fencing until midnight after work. Farms are never finished."

Beautiful and fascinating, elk are masters of body language, Win says. "They communicate in ways you can't imagine. Bugling is somewhere

Veterinarian Terry Church is fond of Hope, a bottle-fed elk
he raised on Canadian Rocky Mountain Ranch near DeWinton.

between whistling and screaming, but they have eloquent grunts and
clicking, puffing—all sorts of vocabulary."

Elk are easier on the land than horses or cattle, although they rub on
trees and make a mess of bush, but their hoofprint isn't as heavy or as
deep, and they eat high-protein leaf litter (fallen leaves), which livestock
won't touch.

Once in the early years, Win delivered a calf whose mother later
rejected her. "When I pulled her free of her mother, her leg broke, so she
had a cast on and we bottle-fed her. Named her Pebbles; she lived with

us for eleven years. The animals were so expensive then, I couldn't afford to lose one. But it isn't something I'd do again. The chances of survival are small. I'd let Mother Nature take her course." Pebbles created her own role, helping to calm the herd by coming to the fence to be petted.

In 1994, he left Siemens Electric to embark on several small businesses, including Brew Brothers, a Calgary microbrewery, a hot air balloon venture, and Western Opti-tech, which calibrates and services medical microscopes. "Science, nature, food, wine, schnapps. My one or two good friends. I am drawn to things that are good for the heart," he explains. "I like Alberta's big spaces and I admire the art of crafting, like that good beer we made. I just don't have much time for unreliable people . . . 'You're not the boss of me'—you ever hear that?"

Win's herd now numbers eighty head, mostly sold as meat to several white tablecloth Calgary restaurants and a health food store. There's not enough money in elk velvet these days to justify the pain antler removal causes, and he sells only the odd bull for breeding. "Elk don't bring me any money, but I will keep them—for my heart!—as long as I don't have to bring a sack of money each year."

Now in his early fifties, the former hunter lay down his gun fifteen years ago. "My freezer is full. Why kill a living being if I am not going to eat it?" He's given up the rat race too, in favour of time with his grandkids. "I'm plenty busy, with landscaping our house, raising baby elk, managing my microscope company. There is a timeframe when you are so busy building up your own life, and then you slow down and see the little gaffers in a different light. They like to go fishing with Opa."

Canadian Rocky Mountain Ranch DeWinton
Dr. Terry Church

Watching Terry Church at Canadian Rocky Mountain Ranch (CRMR), it's clear that his work as manager of the ranch and its large animals is more than a job. When Terry affectionately rubs the coat of Tara the bison, or Hope, one of several bottle-fed elk, his soft spot is obvious.

Hope is low in the pecking order of the elk matriarchy, Terry says. "See? She's been bitten." There's always a boss cow, dictated by size, vigour and attitude that cannot be taught—the conviction of being born to lead.

Submissives like Tara are easily identified, their fur coats ruffled and pocked by teeth marks.

Terry's quick to admit that once an animal has a name, it's out of the meat program. "I like how attractive elk are, and how observant. They've needed to be aware of their surroundings for their survival. They're expressive animals, unlike bison. Bison stand and stare; they don't show you anything, they just have the one look. Who knows what evil lurks in the mind of a buffalo?"

Terry joined the ranch in 1999, after retiring from Alberta Agriculture, where he had been involved with value-added processing. "My son John did the initial work here, and his PHD, with elk and bison," Terry says to me as we drive the bumpy lane to the field. "He sold me a bill of goods, that it would be good semi-retirement work. 'Go dabble with the elk and bison a little, Dad,' he said. Ha."

The ranch, owned by restaurateurs Pat and Connie O'Connor and their son, Brad, is located in the rolling hills southwest of Calgary, near DeWinton, and is home to two hundred bison and three hundred and fifty elk. The animals are raised as breeding stock, for their antler velvet and to supply CRMR's urban restaurants and mountain lodges, including Emerald Lake Lodge near Field and Buffalo Mountain Lodge in Banff.

Twenty-five kilometres of two-metre-high fences keep wild animals out. The Cross Conservation land lies less than two miles distant, and Terry has seen several wild elk hoping to add to their harem. Wild elk unwittingly pose a risk to their domesticated cousins. Chronic wasting disease (CWD), found in wild deer and elk, is slowly spreading from Saskatchewan, Colorado and Wyoming.

"How long will it take to reach us? I don't know, but there are lots of wild deer." Terry lays his hand on Hope's neck. "It would be catastrophic. The only policy adopted by the Canadian Food Inspection Agency is to destroy the whole herd and bury them, including the bison and all other ruminants on the farm. CWD is serious, no question, and for elk and deer, it's fatal. That's why we have a closed herd; the last elk brought in was in 1992."

Internal fences separate the domesticated elk into small groups, called "mobs." Breeding females, cows and calves, and bachelor non-breeding males, occupy separate fields. Three breeding bulls visit their harems when rutting season begins in late autumn.

"Testosterone is a curse," Terry says in resignation. "The bulls lose nearly two hundred pounds and are focused on fighting and females." As he speaks, an elk bull bugles from two fields away. Hope, standing near Terry, lifts her head in unison with every other cow in the field, their long noses stretched upward to their necks' utmost reach. "For better vision, their eyes are well around on the sides of their heads," Terry explains. Such placement is a tremendous asset if a wolf or coyote is sneaking up from behind, but it makes elks almost blind to what is right in front of their noses. Terry Church appreciates the irony.

cherrywood-smoked elk loin
on eggplant salad

This meat has a lovely texture if you cook it to medium-rare. Cooking it longer converts it to shoe leather. Brining helps, adding moisture to a naturally lean meat. Leftover smoked game makes an outstanding sandwich, especially when it is thinly sliced. Substitute venison, bison, lamb or beef. Make the eggplant salad up to a month in advance and store it in the fridge to mellow before serving with bread or crackers, grilled or smoked meats. Serves 4.

eggplant salad:
2 large eggplants, peeled and sliced ½ in (1 cm) thick
1–3 Tbsp (15–45 mL) extra virgin olive oil, for brushing
1 spearmint sprig, minced
fruit-infused vinegar to taste
1–2 Tbsp (15–30 mL) honey
½ tsp (2.5 mL) dry-roasted and ground fennel seed
½ tsp (2.5 mL) dry-roasted and ground cumin seed
½ tsp (2.5 mL) anise seed, cracked
1 tsp (5 mL) dry-roasted and cracked coriander seed
2 garlic cloves, minced
½ tsp (2.5 mL) smoked hot paprika (optional)
¼ tsp (1 mL) hot chile flakes, or to taste
freshly cracked black pepper to taste
kosher salt to taste
⅓ cup (80 mL) extra virgin olive oil or cold-pressed organic canola oil
brine:
¼ cup (60 mL) Demerara sugar
¼ cup (60 mL) kosher salt
3 cups (750 mL) cold water
1 tsp (5 mL) juniper berries
4 bay leaves, crumbled
1 whole dried hot chile, crumbled
½ cinnamon stick, broken in pieces
4 whole star anise pods
½ tsp (2.5 mL) whole allspice berries
4–6 whole cloves
½ tsp (2.5 mL) fennel seed
1 Tbsp (15 mL) peppercorns
2 tsp (10 mL) mustard seed
⅓ cup (75 mL) gin

elk loin:
2 lb (1 kg) elk loin
2 Tbsp (30 m) olive oil
1 tsp (5 mL) dried oregano or dried basil
kosher salt and freshly cracked black pepper

Preheat the oven to 400°F (200°C). Line 2 or 3 baking sheets with parchment. Lay eggplant on the trays in shallow layers. Brush or drizzle sparingly with olive oil. Roast uncovered, turning once as the slices brown. Remove eggplant from the oven and slice into finger-size pieces. Season immediately with remaining ingredients except oil, mixing well. Transfer mixture to a glass jar, cover with oil and fit with a snug lid. Refrigerate for up to a month to mellow.

Combine the brine ingredients in a pot. Bring to a boil and simmer briefly, about 10 minutes. Cool. Immerse the elk in the cooled brine. Let stand 12 hours in the fridge, turning from time to time. Soak your preferred hardwood or fruitwood (cherrywood is lovely!) chunks in cold water for at least an hour or until thoroughly saturated. Remove the meat from the brine and pat dry. Discard the brine. Rub the meat with oil, and lightly dust with oregano or basil and a sparse sprinkle of salt and pepper.

Turn on the grill, setting the temperatures to high on one side of the grill and low on the other. Put handfuls of wet woodchips into a flat little tin, such as an old cake or bread pan, and put the tin on the cooler side of your grill, directly on the coals or heating element. When they begin to smoke, close the lid, and let a good head of smoke develop, about 10 minutes. Place the meat on the hot side of the grill and cover the grill. Grill for 10 minutes for medium-rare, turning once halfway through. Remove the meat from the grill, let rest 10 minutes, then carve thinly against the grain. Serve with eggplant salad.

In one of his rare public appearances, Pincher Creek smoker Joe Cunningham seasons his smoked trout at the Calgary Stampede.

f is for fish

Cunningham's Scotch Cold Smoking Pincher Creek
·Joe Cunningham

A drummer and percussionist confessing to doing things "offbeat" is akin to a dancer saying he can't remember how to waltz. Now in his mid-fifties, Joe Cunningham's drum kit dominates the middle of his living room, and his out-of-sync lifestyle is just a fact. Joe's second CD, *World Weary*, a blend of R&B with jazzy tones, was an indie release in 2010. His third came out in 2011.

Joe grew up in Nova Scotia, where he witnessed the destruction of the Atlantic salmon and cod fishery. He founded a courier company, then a diaper service, while his daughter, Sidney, was a baby in Halifax.

Joe came to Calgary "on spec" in 1991. His wife, Tony Payzant, and Sidney followed a year later. The marriage ended in 1992, and Joe went back to university and did two degrees in social work.

"By 2001, I was fed up with the city and wanted to quit social work," Joe says. "I just wanted to get out of Dodge, but thought I'd start a small business." He moved to Pincher Creek in 2001 with his second wife, Janice Day, whom he calls the practical and organizing force in his life. As a kid, Joe had a dream of farming mussels, an idea that never took root. "I'm a bit obsessive-compulsive," Joe says wryly, "and what interests me isn't practical stuff. I'd never be a plumber or a printer. Aquaculture appealed, but I couldn't do that in Pincher Creek. Seafood in Alberta sucks. So maybe . . ." Joe settled on smoking trout, Janice became the town librarian and Sidney went to university in Calgary.

Joe knew very little about fish politics, but he did some homework. The issues around seafood are complex. "It's never as simple as purchasing wild over farmed." He set up a smokehouse, cold-smoking char, salmon and trout, turning the trimmings into pâté. He now sources fish from

Farmed tilapia is usually raised in a symbiotic environment where its waste water is used to nourish other species.

Wild West Steelhead, located on Lake Diefenbaker, south of Saskatoon. Joe properly calls his fish rainbow trout—they come from fresh water, and "steelhead" is actually the name applied to ocean-going trout.

Joe has gravitated to work that relies on slowness. "I make mistakes going fast to keep up with the world," he says, speaking in deliberate syllables. "I like to work and think slow." Cold-smoking is a two-day process, with little variation and its fair share of drudgery. The reward, Joe says, is as satisfying as making music, or fly-fishing, his other passion.

Joe's learned he'd rather fish than talk. During a brief tenure in 2005 at a booth at the Calgary farmers' market, he had to decide if his time was best spent making his expensive luxury product, or selling it at the market. Either way, he was spending his money on staff. There were too many variables in the smoking process to adequately train another person, he decided. It made sense to selectively wholesale, so his contact with other people is now limited to monthly trips to Calgary, delivering to several specialty food stores.

Two photos hang on a wall in Joe's home. One is of the serene Oldman River, and the other is the chaotic floor of the New York stock exchange. "You can never master fly-fishing," Joe says. "When I fish, I become completely absorbed, the way a human should be. The juxtaposition for me is men in suits, frantically sticking their arms up in the air. What the hell are they doing?"

Joe finally figured out that he was doing something unique in Alberta. "Everyone has things in them that are real strengths," he says quietly, slicing his fish into samples at the Calgary Stampede. "We try to shove people into little boxes, and make a society of cogs and wheels. That's too tough for me. I've never been good at—or happy—having a job. When you're young, there's this pressure to fit in and be normal. Well, I tried, but was never very happy doing it. I am good at simple things, fish and music. I'll keep on doing at least one of them, and now I'm looking for someone to carry on with the smoking."

Greenview Aqua-Farm Delacour
Yan Qian

For years, while eating dim sum with my sons in Chinatown, I watched tilapia swimming circles in their aquaria. In 2001, I realized that *someone* had to be raising them nearby. I contacted Victoria Page to finagle a fish farm stop on the third annual Foodie Tootle bus tour.

When we arrived at Greenview Aqua-Farm, near Delacour, east of Calgary, a busful of Calgarians walked the elevated plank floors beside several tanks and watched Victoria feed tilapia destined for the tanks and platters of Chinatown. Victoria described how the fish tanks' waste water was cleaned and re-oxygenated by the plants in the greenhouse. Adjacent wetlands filtered any remaining waste water while providing homes to eighteen varieties of birds. One small tank seethed with eels, vividly reminding me of scenes from *The Tin Drum*, the 1979 movie based on the Gunter Grass novel.

Fish farming is not new. Over the ages, water-covered rice paddies in Asia have harboured tilapia and carp. After the rice comes in, the fish are harvested, a farm practice that encourages dietary diversity and health, environmental symbiosis and financial stability. The majority of prairie aquaculture is trout fingerlings to stock ponds and dugouts, and tilapia, or "Nile perch," a warm-water fish originally from Africa. Tilapia need air, moving water and a balmy 23°C setting for their water. A power failure would cut off the air supply to the fish tank and the fish would suffocate within twenty minutes. In aquaculture, escape is also a primary concern. Escaping fish may spread disease or take over a waterway without natural predators to keep them in check. At Greenview, surrounded by a sea of prairie, risk of a breakout is minimal.

Greenview and its marketing arm, Flatlander Fish, began in 1994, springing from the longing of two Chinese-Canadian businessmen for the fresh fish they had enjoyed as children in China. Tang Lee and Bill Ho teamed up with Darwin Monita, a water toxicologist, and hired Victoria, who was pursuing her MA in education. They set up a closed-loop system that kept the fish and land fed.

Current Greenview owner Yan Qian no longer has a greenhouse, but the waste from the fish is used as fertilizer on a neighbour's farm. He sells ten thousand pounds of live tilapia to Asian markets in Calgary and Edmonton.

"Caucasians don't care if their fish is live or fresh," Yan, who shuns the spotlight, tells me. He isn't completely accurate, although human greed and short-sightedness are responsible for the abysmal state of the world's fishery and the depletion of the oceans. In September 2010, the

United Nations' Food and Agriculture Organization (FAO) drew up guidelines to govern food safety, environmental and social issues related to aquaculture.

Although the world's wild fishery is in decline, the global appetite for fish is not. Mike McDermid, manager of Vancouver Aquarium's Ocean Wise program, estimates that in Canada, over one hundred inland and ocean farms sustainably produce tilapia, trout, char, halibut and a variety of shellfish.

In southern Alberta, near Nobleford, Klaas Dentoom began farming tilapia in 2007, and ships his catch live to Vancouver's Asian markets. Mark McNaughton also raises tilapia, at MDM Aqua Farms, east of Three Hills. He too wholesales to Asian markets. "Fish farming isn't so different from conventional farming," Mark says, "just in a different medium." He converted a hog barn into a home for thirty tanks filled with tilapia. Fish manure enriches his fields, and he raises aquaponic tomatoes and lettuces in an adjacent greenhouse for sale at the farmers' market. Mark has spent over $100,000 on equipment since beginning in 1999. He grumbles when he hears people call aquaculture a get-rich-quick scheme. "It's just as time- and labour-intensive as any other livestock."

Swift Aquaculture Ponoka
Bruce and MaryLou Swift

After seeing open-water salmon aquaculture pens for myself in the Inside Passage near Campbell River on Vancouver Island, it is a relief to speak with Bruce Swift. He's an Ontario boy who moved to the west coast when he realized he wanted to farm salmon.

The owner of Swift Aquaculture has been "in coho" since 1986, when he wrote his UBC animal sciences Master's thesis on animal breeding. Bruce's first fish farm was in Sechelt, on the Sunshine Coast, as the first contract fish grower for BC Packers. He moved to Abbotsford when algae bloom wiped out 100 per cent of his neighbour's fish.

Now in his mid-fifties, Bruce has coho salmon eggs in quarantine at Lethbridge College's aquaculture centre as a preliminary step towards developing an aquaculture farm in central Alberta. "Alberta is home. My wife, MaryLou, works at a barley breeding lab in Lacombe, and we have five acres in Ponoka," Bruce explains. "When we get Alberta coho established, anything we do will have her nutrition work as an integral part, for knowledge, health and for a sustainable business."

The couple received health certification for the eggs in 2010. The next step will be a commercial trial.

Bruce's current four-acre site in Agassiz, BC began inauspiciously in 2004 when not a single Vancouver restaurant would accept his farmed coho. He sent over twelve hundred fish to a mink farm as animal food. He now grows several species in a closed-containment system of inland tanks, in a symbiotic system known as integrated multi-trophic aquaculture (IMTA). It's a "closed" food chain: salmon's solid waste nurtures vegetable fields and waste water grows greens, including wasabi and watercress; the algae become a feed supplement for the farm's freshwater crayfish. Bruce supplies several Vancouver restaurants, but mostly, he raises salmon brood stock.

"Aquaponics is a no-brainer," he says. "With the amount of water you are re-circulating, there are always surplus nutrients. Why not try to grow something else with it?" He keeps the fish small, under two kilograms, and infrequent handling keeps stress levels for the fish low.

In Alberta, Bruce faces different water conditions than BC's alkaline levels and salt content. He's paying close attention to locale for other reasons as well. "I'm big into local food," he says. "I want to be close to Calgary or Edmonton, ideally in the middle, so we can deal with a small local chain or retailer."

"We love growing fish. It's a good diversification product for agriculture." He's unsure if he and his wife will take on the Alberta location themselves. "It depends on our son." Eric, in his early twenties, is currently managing the Agassiz farm.

Bruce envisions eventually offering a partnership package—fry, expertise, technology, and marketing—to interested prairie farmers. "Farmers have backup generators, labourers, spare buildings to convert for tanks. Aquaculture could be mainstream in five years. It's just farming. Farmers don't waste nutrients." Aquaculture has potential for much more than food, he says. "There's a lot of research going on—heat exchangers, methane generators. In the next ten years, we will see a lot of shifts."

Farming fish

Opinions on farmed fish have drastically changed, thanks to educational work by organizations like the Vancouver Aquarium's Ocean Wise program and the biennial Canadian Chefs' Congress.

Fish farms fall into two categories: inland tanks and open-water pens. Generally speaking, inland farms do a better job of controlling problems than open-water pens, although black tiger prawns raised in Southeast Asian ponds are a glaring negative example. Data recorded by the Trade Environment Database (TED) indicate that coastal prawn ponds the size of football fields are dug in areas populated by mangroves. Mangroves, more than one hundred species of trees and shrubs that thrive in coastal salt water, create a nursery-like host environment for a wide array of fish and crustaceans while their roots dissipate wave energy. The ponds, which yield as much as ten thousand pounds of prawns annually, last only two to four years and then are deserted. The mangroves are killed by fecal pollution and the area is often impossible to reclaim. In Thailand alone, three-quarters of the country's million acres of mangroves has been destroyed by shrimp farms.

In open-water pens, population density leads to increased parasites and large amounts of excrement. Escape is a risk, and the formation of associated algae bloom reduces the ocean's oxygen supply. Off-shore pens must be robust to withstand rough seas, and the costs of transporting finished fish are high, in dollars and in fuel.

Fish farming is experiencing a global renaissance. In September, 2010, the United Nations' Food and Agriculture Organization (FAO) drew up guidelines to govern food safety, environmental and social issues related to aquaculture. Mike McDermid, manager of the Vancouver Aquarium's Ocean Wise program, says, "We know we cannot support the demand [for seafood] without aquaculture. We need to sustainably farm fish."

Aqua-farmer Bruce Swift believes that "coho salmon has the potential to be the aquatic version of broiler chicken." The species grows rapidly—two years is the time span of a generation—and is virus-resistant, but what farmed fish eat is a hot button. Some farmers convert their herbivores into piscavores by giving them fish meal for faster growth. Does this make them a worse ecological choice than fish from farther away that were raised on a sustainable vegetable-based diet? Dr. George Leonard of Ocean Conservancy, an American non-profit conservation organization, speaking in September 2010 on the CBC radio show *Quirks and Quarks*, says, "The feed required by piscavores (fish-eaters) like salmon is the Achilles heel of the aquaculture system." Salmon require pound-for-pound amounts of fish meal, fish oil and smaller fish. That means fishing or farming fish to feed the fish. One-third of the world's fish is farmed, so what to feed the fish is becoming a bigger question.

Some food, like Peter's fishes in Galilee, is hauled from the water. But Galilee doesn't freeze, and Peter never had to wear four layers of thermals in -30°C. Kellen Wickenheiser, who fishes commercially for whitefish in southern Alberta, dons two pairs of Stanfields under his jeans and overalls. Then he puts a forty-five-centimetre auger through the ice on man-made reservoirs that are nearly fifty kilometres across. "That's not even a bay up north in Lake Athabasca," he tells me. "In winter, after we drill a hole, we put a jigger in, send the jigger out one hundred metres and drill another hole. Then you hook a net between the two holes and pull the net in."

Kellen fishes in fall and winter because ice is too expensive in summertime. "In winter, we have all the ice we need." In fall 2010, Kellen and his three-man crew pulled in one hundred thousand pounds in twelve days, mostly three-year-old whitefish, with some pike and walleye.

It's a tough, physical job. Kellen, in his early fifties, is six feet tall, over two hundred pounds. He's worked outside most of his adult life, including twenty-six years on drilling rigs, and figures he's good for another thirty years of fishing. "I think men especially are physical beings and need to do hard work," he says. "This is tougher work than the rigs. After I bought this licence, one of my derrick men came to work for me on the boats. When we got to shore, he said, 'holy crow, you work hard for your money!'"

A large percentage of his clients are Hutterite colonies; a fraction ends up in a few stores; and some Kellen smokes over maple wood and sells privately. If the fish isn't bespoke, it goes on ice to Freshwater Fish Marketing Corporation on Great Slave Lake, and on to the processing plant in Winnipeg. There, if it's autumn, the roe are removed for shipping as whitefish caviar to Europe and Asia.

Whitefish have always been prized for eating. "Until the Depression," Kellen says, "the CPR's 'silk trains' carried bales of Japanese raw silk from Vancouver to New York's mills. Those trains rushed across the country in four days, but they stopped to stock up on whitefish, for the crew, see, and to sell at the other end, that's what the old-timers tell me." Whitefish's popularity has waned. The province used to issue thirty-five licences. "Now it's just me, another guy my age, and two old guys."

Each summer, Kellen gardens. He raises Russian Blue potatoes, a half-acre of Russian Redneck garlic, hemp and flax. His marketing and distribution system is thoroughly contemporary. "It was my wife's idea to sell my garlic on Kijiji," he says. "It goes quick." Internet sales offer a modern counterbalance to the primeval nature of fishing, and its ancient symbol, the net.

smoked fish chowder with chives

Smoked fish, a staple on the Prairies, adds interest and flavour to soup of any sort. The type of tea you choose as a smoking medium will affect the flavour, so choose a hearty smoky tea like Lapsang Souchong for the smokiest flavour. A milder flavour will result from using a bright Darjeeling. Remember as you chop your vegetables that smaller pieces cook more quickly. Serve this rich soup in small portions. Use tilapia, trout, steelhead, salmon or any fish you love. Serves 8–10.

soup:
2 slices smoked side bacon, slivered
1 onion, minced
2 carrots, diced
1 leek, sliced
1 celery stalk, diced
6 garlic cloves, minced
½ red pepper, diced
1 tsp (5 mL) dried thyme
1 bay leaf
½ cup (125 mL) dry white wine
2 cups (500 mL) diced potatoes
4 cups (1 L) chicken or fish stock
1–2 Tbsp (15–30 mL) cornstarch dissolved in
    ¼ cup (60 mL) cold water
¼ cup (60 mL) heavy cream
kosher salt and hot chile flakes to taste
smoked fish:
1 lb (450 g) fish, diced
2 Tbsp (30 mL) black tea
2 Tbsp (30 mL) white sugar
2 Tbsp (30 mL) uncooked rice
garnishes:
3 Tbsp (45 mL) chives or green onions, minced
1 lemon, zest only

For the soup: Slice bacon finely and sweat it in a heavy-bottomed stockpot. Discard any extra fat after it renders out, leaving 1 tablespoon (15 millilitres) in the pot. Add diced and sliced vegetables and cook them without colouring, adding small amounts of water as needed. When vegetables are tender, stir in thyme, bay leaf and wine. Bring to a boil, then add potatoes and stock. Simmer, covered, until potatoes are tender. Return to the boil and stir in cornstarch dissolved in cold water. Return briefly to the boil to allow the starch to thicken, adding more dissolved cornstarch for a thicker soup, then add heavy cream, salt and hot chile flakes.

To smoke the fish: Line the bottom of a wok or heavy-bottomed pot with a piece of aluminum foil, about 6 inches (15 centimetres) square. Measure sugar, tea and rice onto the foil and mix it around. Place a wire rack in the wok or pot, positioning it above but not touching the tea mixture. Lay fish on the rack in a single layer. Cover snugly with foil, then with a close-fitting lid. Put the wok on its ring, over high heat and cook, covered, until the fish is cooked through, about 5 minutes. (Larger amounts will take longer.) When done, the fish will be brown, opaque and firm to the touch.

To serve: Ladle the soup into heated soup plates. Top each portion with several cubes of fish, and a sprinkle of chives or green onions, and lemon zest.

Grilling beef over open flames in the back country of Alberta.

g is for grass-fed beef

Trail's End Ranch Nanton
Linda Loree, and Rachel and Tyler Herbert

The Indian summer sunlight is lengthening across the Porcupine Hills as I head south and west from Nanton, toward the Chain Lakes Provincial Park, looking for Trail's End Ranch. Where the road dwindles to a trail, I meet Linda Loree. We park high above the homestead and walk in. Linda looks like the quintessential old-time rancher: tall, lean, weathered, wearing a timeless broad-brimmed hat. She has taken up the family land and raises cattle with her daughter, Rachel, an experienced horsewoman, her son-in-law, Tyler Herbert, and two young grandchildren.

As we walk down the hill, I want to ask about the death of her son, Nathan, in southern Afghanistan in 2007 at age twenty-four, how it changed things, if his death is why she has gone back to her roots. But I'm hesitant, reluctant to intrude, even though Nathan was my eldest son's best friend. I don't need to ask. Linda brings up her lost boy without tears. Her elder grandchild is named William Nathan, she says, as we watch the little boy rummage through the short grass next to the fenceline. "It makes for a nice symmetry," she adds, "to have the fifth generation on this ranch. Nathan would be glad."

Linda's grandfather Fred Ings ran cattle on this range in the late 1800s, and started several others in the area. Fred's chronicle, *Before the Fences*, written in 1936, was first published in 1980. After he died in the 1930s, the house was used by his widow, Edith, as a B&B and guest ranch, mostly for visiting Brits.

"My great-grandmother came west in 1909, from an Ontario family of privilege," Rachel takes up the family story. She took a bad fall from her horse in 2002, and retreated to the ranch with her puppy. "I stayed into the winter and commuted to university," she recounts. "For me, with

Linda Loree, right, ranches with her daughter, Rachel, left, son-in-law, Tyler Herbert, and their youngsters, Avery and William, in the Porcupine Hills southwest of Calgary.

an English degree on the Romantic poets, it's always been the romantic element of ranching that appealed. My Master's is in natural history, on how people have related to the land here. So anything incompatible is unappealing, for example, feedlots. I feel that grass-fed ties into that old-fashioned way of raising beef." Rachel's research into ranching history taught her that ruminants like cattle do not thrive on grain. An animal must be gaining weight when it is processed, or the meat will be tough, so in the winter, they supplement the yearlings with a sprouted barley malt pellet. This by-product from brewing acts like forage, and doesn't cause acidosis in a cow's first stomach, the rumen.

Tyler was cowboying when she met him in 2003. A year later, they married and bought a quarter section on the flat land south of Nanton. "A coulee runs through it, bumpy and hilly, it's never been broken up, even though it is surrounded by farmland," Rachel says, settling down to nurse her daughter, Avery Edith. "We were lucky to get this little piece of grass. Keeping land in grass is a better environmental long-term plan than large commercial chemical based farming. More grass means more food from grass."

They sell direct, via the Internet, live year-round on their ranch, and visit the Trail's End house in the summertime for getaways or when fencing chores call. Income from their herd of seventy cattle is supplemented by Ty's off-ranch work in the oil fields. Rachel hopes he will eventually work closer to home as a general contractor.

"His job is how we can afford it. It's a choice of how we want to live as much as how we want to eat. They go hand in hand. I need this. I'm trying to develop as a whole person with respect for the land. We want our kids to be physical and active, to see real things." Things her brother Nathan didn't live to return home to, but learned to value as a boy, playing on the sunny hillside where Rachel sits, nursing her baby, while William Nathan runs around beside the fence.

Sun to Earth Farm Castor
Kathleen Charpentier and Richard Griebel

The woman I regard as eternally joyous is captured in a moment on a bridge in Torino in 2008. When she saw me, Kathleen flung her arms into the air, stood poised, exultant, fingers open, arms wide. She was approaching the site hosting Slow Food's Terra Madre, and nothing but a full-body salute would suffice to express her pleasure. Several years later, in a great rush of words, exuberant, gleaming as brightly as her teeth, Kathleen Charpentier leans forward and speaks, her hands as animated as her face. "Richard and I sleep in the bedroom that he grew up in."

Kathleen and her husband, Richard Griebel, work and live on the family farm, and have taken a deliberate step back in how they tend their animal charges. Back, they concur, but not backwards. "We did agriculture as it should be done, then and now," Kathleen says. They take turns, sometimes speaking at the same time, finishing each other's sentences, the words rolling and jumping into each other like lambs gamboling in the field outside the house, recounting Richard's childhood in a large farm family, tending a huge garden, making sauerkraut, rendering lard, milking shorthorns, keeping chickens. "Every animal was on grass. Then I went to Olds College in 1976 and got educated." Richard's voice changes. "I learned the corporate industrial model of farming. Everything was based around production and quantity."

Kathleen Charpentier of Sun to Earth Farm celebrates her arrival in Torino, Italy, to attend Slow Food's Terra Madre gathering in 2008.

He stops, looks through his pages of notes, quotes Carlo Petrini, the founder of Slow Food. "Be united in respect for the earth and our conscience. Live intensely." Carlito, Slow Food's passionate speechmaker and evangelist, could have been talking directly—and only—to this couple, who burn with the same fervour.

Kathleen interjects, "If we can create a parallel system, we can have a place and way to feed people for when oil goes up. We don't think, in this country, that we will be hungry, but we are living under an illusion with cheap fossil fuel. Things can change so fast . . ." She and Richard have taken Petrini's leadership into their lives and internalized it. Their lives revolve around their deeply held belief in the sanctity of soil, and the honour in treating animals with grace and dignity.

Kathleen and Richard tend sixteen quarters of land in the dry east-central region of Alberta where shortgrass can still be found, where it takes twenty acres to keep a cow, where the grass grows slowly. "We are always pushing," Richard continues. "We have broken our contract with nature. Putting animals in confinement barns in feedlots, dumping chemicals on our souls. I don't know how we got so lost."

On their farm, they keep six hundred to eight hundred meat chickens, a milk cow for the house, forty laying hens, thirty Rouen and Pekin ducks, forty Toulouse and Embden geese, forty sheep and their sixty lambs, one hundred and twenty Dexter and Galloway cows and their calves, one hundred and twenty yearling cattle, and fifteen to twenty pigs. All the animals graze, in a graceful arc, from one field to the next, a layered mosaic of multiple species.

In 2011, Richard was diagnosed with a virulent form of thyroid cancer. They took on a partner, twentysomething Blake Hall, and their adult son Ian returned to the farm. "Our biggest dream is when we leave, we have five families making a living from our farm. We have to get bigger, but in numbers of people on the land, being paid fairly for good, clean food."

"I really want children eating our food," Kathleen says wistfully. They spend some time quietly talking, analyzing the omega-3 value of grass-fed meat, then say what they believe to be right. "We have a beautiful life," Kathleen says, utterly sincere. "We nurture our land and plants and animals, but also our families and our friends. We attach honour and sanctity to life; it is missing in our culture."

Hoven Farms Eckville
Tim and Lori Hoven

"All business comes down to relationships. That means that people are the most important thing," Tim Hoven says. He stands in his bustling booth in the centre of Calgary's newest farmers' market as we talk. Tim is one of the founders of Kingsland Farmers' Market, which opened in July 2010 on Macleod Trail South, one of Calgary's busiest roadways.

Tim has nurtured good relationships with his customers since Hoven Farms began its direct sales in 1997. He credits his dad with the family's current steady position. "My dad is very long-sighted, it was 'everything natural' with him. He was right in 1997. It meant that when BSE hit, we had our feet firmly planted in direct marketing. It kept the farm above water, and now, we are thriving."

Tim greets another vendor before turning back to me. "If a farmer is not direct marketing, it's almost impossible to have a profitable thriving farm. I'm not interested in doing the export market route, because my land can only support so much beef without becoming a big intensive farming operation, effectively a feedlot. Calgary is big enough to take all the beef I can grow."

Tim, his wife, Lori, and their eight kids raise their Angus-cross beef near Eckville. "As far as I'm concerned, good management goes a lot farther than any breed," he says of his herd. About four hundred and sixty animals each year are taken to a local slaughterhouse, where they are humanely killed and processed. Tim pays special attention to the weather when feeding. "At minus-35 with wind chill, you can't get enough weight gain on an animal without grain. The outside fat will be fine, but the steaks are chewy, so at the first cold winter breath, I stop grass finishing."

The Hoven management style has won them many friends and admirers, including outspoken Calgary chef and local foods supporter Wade Sirois, who accompanied a calf whose birth he had witnessed to the slaughterhouse. "I came away with a feeling of reverence," Wade says, "for what and how Tim does things, for life, for knowing where it all happens, and how, in whose hands."

With four of their eight kids currently being home-schooled, Lori sacrifices any free time to nurture her brood. "Kids need a good foundation, that time with Mom and Dad. We eat lunch together; their dad is at home except for setting up the market. It gives them life skills." A lot of family time is spent on food. Lori says, "Our kids are learning to grow, maintain and harvest the garden, to use machinery, and how to cook. There are so many of us; Mommy can't do everything."

Meals are simple, their own beef and garden produce, supplemented by local vegetables. "We are a meat and potatoes family, and we have chosen to not be bound to eating a huge array of different foods. We keep it simple so we can manage meals."

The house was raised by Tim's grandfather, and was home to Tim's parents before Tim and Lori took it on. "Our kids are the fourth generation to live here. We have a vision for them," says Lori. Like any mother, she hopes her kids will "spread their wings, gather many experiences, then be back."

Her husband wryly says, "Marketing has taken the forefront, and it's lots of effort. I don't mind doing that so I can take a step back in a bit." He adds, "What I love to do on a beautiful summer day is go check my fences. No cellphone, just the sun beating down. The cows chewing is a beautiful sound."

The Producers of the Diamond Willow Range and Diamond Willow Organics Ltd. Pincher Creek
Keith and Bev Everts

If you want to see a rancher, you have to visit the ranch. So I put on my boots and hit the high country. Despite the greyness of the day, snow flying into my windshield as I drive, I am happy to be in the foothills, but I am relieved when I finally see the carved wooden sign that flags Francis and Bonnie Gardiner's Mount Sentinel Ranch.

Inside the old house, the day unfolds like an oil painting from another era. The ranchers stand close to the wood-burning stove, capturing its heat on their weathered skin, then move to the acre-long kitchen table. Keith Everts of Stillridge Ranch, his long blond braid fading to grey, leans his elbows on the table beside his colleagues and talks about biofuel. Francis Gardiner eats quietly, then retreats to his front room to read about wolves. Later, I bundle up and find him outside, watching a calf in the snow.

This seven-family ranching corporation raises certified organic beef in the foothills south of Calgary, on a total of sixty-three thousand acres of land from Longview to Pincher Creek. They've been the model for co-operative meat marketing since they joined forces in the mid-'90s. Francis and Bonnie Gardiner own Mount Sentinel Ranch; Cass and Jamie Freeman have Freeman Ranch; Larry and Jan Frith work Frith Ranch; Keith and Bev Everts own Stillridge Ranch; Janet Main, Charlie Straessle, Mac Main, and their families run MX Ranch; Salix Enterprises is the home of Bill and Carol Elton; and Ketaorati Ranch is run by Norman and Hilah Simmons. Some of the families involved have been ranching for over a century.

The issues facing ranchers are as diverse as this gathering. Keith recounts that the word "organic" was a hard sell in the 1990s, when he

Half the members of the Diamond Willow gang gather with chef Scott Pohorelic in the kitchen of Francis and Bonnie Gardiner's Mount Sentinel Ranch.

broached the idea of joining forces. He and his wife, Bev, the youngest of the original seven, had been hippies on Vancouver Island. Keith's folks had a grain-trucking business in Alberta, so they came back to be near them and to make a living on the land. "In Alberta, in that country, the way to make a living was to raise grain," Bev says. "But after Keith was hospitalized from the effects of chemical sprays, we decided we were not the least bit interested in that lifestyle. Beef was the alternative to grain." It took talking and countless cups of coffee, but ultimately, the others realized that the ideas underlying organics matched their existing ranching practices.

Bev's mom was Olympian swimmer Frances Smith. "Athletics teaches a work ethic and discipline that lends itself well to getting on in the world," Bev says. She raised her three kids on the ranch as active skiers and swimmers. Their daughter and son-in-law have bought Salix Enterprises' herd as Bill Elton slowly retires. (Francis Gardiner died in August 2011.)

"We are still paying for the land, and will until we die," Bev says. "So we don't just ranch; we have other jobs." She believes that many consumers want to know where their food comes from and that it is traditionally raised. "Authenticity is what people want," she says, "a connection." Keith co-founded Southern Alberta Land Trust (SALT), and Janet Main has committed part of her ranch to the Nature Conservancy and to SALT; others have donated land. "It's not new," Bev reiterates. "There are traditional ranchers following old traditional ways, who were organic before there was the word."

TK Ranch Hanna
Dylan and Colleen Biggs

The endangered northern fescue grasslands in east-central Alberta sit squarely in the middle of Special Area Number Two, part of seven million acres of land that were sold to settlers for a pittance in the early 1900s, and taken back for unpaid taxes when the Depression hit. Part of the near-desert, sparsely populated land became TK Ranch, founded by Thomas Koehler Biggs in 1956. Tom was a New York cowboy-culture aficionado who married a local girl, Mary Hallet, daughter of pioneer Jack Hallet. Mary, an artist, painted her landscape, and taught her greenhorn husband how to ranch.

One of their sons, Dylan, stayed on the ranch and became an expert in the humane management of cattle and horses, a skill he presents in seminars across North America. In 1985, he adopted holistic management principles—decision-making based on environmental, social and financial sustainability—in his ranch's land and animal management. Five years later, he met Colleen Nelson, an environmental and animal rights advocate.

Colleen, now in her late forties, had survived stage four cervical cancer in her late teens by adopting a vegan diet instead of chemotherapy. Her complete remission led to her interest in the Weston A. Price Foundation and an ensuing degree in environmental sciences. Meeting Dylan at an environmental meeting caused a complete turnaround in her view of ranchers. She and Dylan began their family in 1991.

"It was an eye-opener for a city girl," Colleen recalls. "When the cattle market took a dive and the bank called, I knew there was something wrong. How can anyone in agriculture sleep at night without knowing what they would get for their crop or animals?" In 1995, they began a grass-fed beef sales program and were certified organic in 1998. Colleen became the ranch's full-time marketer, selling directly to health food stores and chefs.

The 9/11 terrorist attacks of 2001 had a ripple effect on TK Ranch, thousands of miles away. Reservations for fifty thousand hotel rooms in the Banff–Calgary tourist corridor were abruptly cancelled, leading to the termination of orders for twenty-three thousand pastured chickens raised by twelve families who knew the Biggs had solid relationships with restaurants. When they asked Colleen for help, she bought the birds outright, then spent eighteen months selling them off piecemeal. "The outlay nearly broke us," she admits.

Colleen keeps an organic family garden, but she believes that the term "organic" has been compromised by the emergence of factory-style organics, which ignores the issues of carbon footprint or animal welfare. No longer a vegan, she calls herself a steward first and foremost. "The animals we raise deserve to be treated with the utmost respect, including their emotional state when they die." So Dylan accompanies their animals to the abattoir in Duchess, keeping them calm until the moment they are killed. "'Everything should be cheap' is what people think, but you can't put a price on animal welfare."

The Biggs' other passion, the biodiversity and productivity of the grasslands, is accomplished through time-controlled grazing in small fields. The couple's efforts have paid off. "The Special Area board uses our ranch as a benchmark for grasslands productivity," Colleen says.

The ranch's walls are arrayed with industry and environmental awards, and their four daughters are active in their ranch life and meat business. Colleen laughs as she tells me, "The banks still look at what we do as high risk." She has plans for a farmers' market, a demonstration farm, store, restaurant and growers' distribution centre. "Change is a scary word to ranchers. Ranchers like to do what they like to do. Our results have changed people's opinions. We don't stand on a soapbox, that's not how people change their thinking. They see the results."

The Weston A. Price Foundation

The Weston A. Price Foundation was founded in 1939 by nutrition pioneer Dr. Weston Price. This non-profit charity supports Dr. Price's research that humans achieve perfect physical form and perfect health only when they consume nutrient-dense whole foods and fat-soluble activators found exclusively in animal fats. The foundation aims to restore nutrient-dense foods to the human diet by accurate nutrition instruction, organic and biodynamic farming, pastured livestock, community supported agriculture, clear and honest labelling, prepared parenting and universal access to clean, certified raw milk.

honey-herb-cured alberta beef or bison steak with spiced honey gastrique

Steak, with a spicy sweet-and-sour honey sauce. Simple. Scrumptious. A gastrique is a fat-free sauce traditionally used to accompany fish or other roasted meats. It should be acidic, not sweet, so add lemon at the end to balance it. Serve hot on grilled or roasted foods. Serves 6.

steak and rub:
6 beef or bison tenderloins, rib-eye, strip loin or other good grilling cut
1 Tbsp (15 mL) cracked or ground fennel seed
1 Tbsp (15 mL) minced fresh rosemary
1 Tbsp (15 mL) minced fresh thyme
5 Tbsp (60 mL) grated fresh garlic
4 Tbsp (60 mL) liquid honey
2–4 Tbsp (30–60 mL) olive oil or cold-pressed organic canola oil
kosher salt and freshly cracked pepper to taste
gastrique:
6 Tbsp (90 mL) wildflower honey
4 garlic cloves, minced
⅔ cup (180 mL) mead, fruit wine or white wine
6–8 Tbsp (90–125 mL) fruit-infused or mead vinegar
½ tsp (2.5 mL) cracked peppercorns
3 whole star anise
¼ tsp (1 mL) cracked fennel or anise seed
1 shallot, minced
1 Tbsp (15 mL) minced fresh thyme
1–2 lemons, juice and zest
kosher salt to taste

Smear steak with all other ingredients. Let stand.

In a shallow sauté pan over high heat, melt honey and allow it to caramelize. Add garlic and stir until garlic colours, then slowly add wine and reduce by at least half the volume. Add vinegar, spices and shallot. Reduce again by one-third. Add thyme, lemon juice and zest. Pick out and discard star anise. Season to taste.

Preheat the grill to medium-high. Grill steaks to medium rare, or your preferred degree of doneness, turning once. Let rest at least 5 minutes, then carve. Serve with a drizzle of gastrique.

h is for honey

Kemp Honey High Prairie
Rachel and Ryan Kemp

It isn't every kid whose house is visited by a queen, or queen-to-be. For Colten and Conner Kemp, queens are a matter of course. The young boys live on a farm near Grande Prairie, and were born into a family of beekeepers. The queen who comes calling is the new queen bee, ordered from as far away as Hawaii, who arrives in a small box, waiting for her rival to be dethroned before she assumes her role in the hive.

The boys' dad, Ryan, and uncle, Tyson, manage about fifteen hundred hives between them. "Ryan has been around bees his entire life; it's a family thing, going back to his grandfather and great-grandmother in Ontario," Ryan's wife, Rachel, says. "Sometimes he doesn't wear a veil or gloves or even the suit when he's working." The thought of going unprotected in the presence of bees is not her idea of fun. But even without the potentially painful animosity of bees taking issue with being robbed, it's physically demanding work to remove frames from the hives. A hive holds ten wooden frames. When filled with honey and capped with wax, each frame can weigh up to ten pounds. The men take the frames to the honey house on their parents' farm a mile away. There, they cut the wax caps off and load the frames into the extractors, large centrifuges mounted in closed tubs. The whirling extractors spin the liquid honey out of the wax honeycomb. As members of Beemaid Co-operative, 99 per cent of their honey is shipped to Spruce Grove to be packaged. The rest is sold in the family store, and some is used by Rachel for her own contribution to the family business.

In 2000, Rachel researched making a natural line of bodycare products using bee-made materials. Wax, propolis and honey go into lip balm, scented skin lotions, bee butters, bath salts and hand salve. "Everything I

Art and Cherie Andres don their business attire to display their vinegar and mead on honeybee hives at Chinook Honey Co.

make smells sweet, like honey," she says, "I really noticed it when I began, but I live in the middle of it, and I can't smell it anymore."

Bees' well-travelled feet drop pollen into flowers and nectar into hives, but in the larger sense, bees play a pivotal role. As pollinators of plants, they unknowingly transmit life, the creative energy that circulates the wisdom of species diversity and fecundity.

Paradoxically, these insects with the busy wings and brief lifespans have fascinated humanity for millennia. The fascination runs deeper than our taste for sweet things. Ancient history, from the kitchen to the church, is rife with references to honey, bees and honeybees. During the Middle Ages, the English tradition of honeymoons purportedly began with the gifting of honey wine, or mead, to the newly wedded couple so things would begin as sweetly as possible.

"Honey sustains us most years," Rachel says. "Honey pays the bills, just like crops in other farming. But you don't farm bees. You manage them," Of course, to beekeepers' kids like Conner and Colten, managing a life based on honey is as routine as a visit by a queen.

Inside the hive

In the natural matriarchal course of events in a hive, a queen bee's rule lasts for two or three years. The queen's life is devoted to the laying of eggs, after a single venture into the outside world to mate. Thereafter, she lays up to two thousand eggs per day. When her productivity declines, several of her eggs are fed royal jelly. The first to hatch begins life with a royal coup d'état, mercilessly killing any contenders for the crown, including her mother.

A glass-covered observation frame is one way to view the society of bees. Worker bees, all female, use body language as articulately as GPS to inform their hive-mates of the location of pollen and nectar awaiting attention, their gyrations invoking the angle of the sun as a key piece of the travel puzzle. They are the guards, the nurses, the fetchers of pollen and nectar, and the builders of the hive itself. In a hive of sixty thousand occupants, all but six hundred drones are female. Drones have a short but sweet life with a limited function: procreation. After the speediest mate with the queen on her single foray outside, they die.

Chinook Honey Co., Chinook Arch Meadery and Chinook Vinegar Works Okotoks

Art and Cherie Andrews, and Pamela Vipond

What began as a stress-busting hobby for airline pilot Art Andrews has become a full-time business with his wife, Cherie. The couple tend two hundred and fifty hives at their foothills farm just south of Okotoks and make an array of honey-based products, including mead and vinegar. Art thinks the bee venom transmitted by countless stings eases his arthritis, but he's still quick to lead me indoors when the bees are cranky during my late autumn visit.

After years of lobbying the Alberta government, Art and Cherie opened the province's second cottage meadery in 2008. Mead relies on time to transform three ingredients—water, yeast and honey—into greatness. "It's simple, but you have to make mistakes to make mead well," Art admits. Three pounds of honey produce a gallon of medium-sweet mead. He is experimenting with traditional mead along with melomel, the old English name for fruit-flavoured mead, and methaglin, mead infused with herbs and spices like nutmeg or ginger. Blackcurrants, cherries and raspberries have made Art's best melomel to date. Good mead can age up to fifteen

Smoky, dee's half-Siamese cat, relies on mead vinegar made by Chinook Vinegar Works to season her salads.

years, in wooden barrels or in the bottle. Art uses two one-hundred-litre oak barrels built by a cooper in Oliver, BC, the new heart of the southern Okanagan wine industry, to age some of his mead.

Standing beside the extractors, Art launches into his "crash course on bees." Bees, hugely important for crop pollination, are at risk from crop spraying that doesn't differentiate between "good" and "bad" bugs, and from CCD, or colony collapse disorder, which has decimated the bee population in North America. (*Today's Diet and Nutrition* suggested in a 2011 issue that gardeners consider planting bee balm, thyme, mint, Russian sage and sunflowers to help the dwindling population.) One bee produces one-quarter teaspoon of honey in its lifetime, and can fly five to seven miles to collect pollen, but studies show that a hive's net production diminishes when bees fly farther than four miles. In Alberta, a good hive will produce in excess of one hundred pounds of honey annually, and even more in the Peace Country, where long hours of daylight and wide swathes of fields make northern bees the province's most productive.

Cherie optimistically hopes to see an Albertan fruit winery and meadery agritourism route similar to those of the Okanagan Valley and Niagara Peninsula. Patience can perhaps be learned from handling bees, and bureaucrats.

The on-farm shop carries honey, apitherapy products including propolis and royal jelly, bodycare products based on honey, and honey ice cream, made in High River with local blackcurrants.

Art and Cherie gave a carboy of mead to Black Diamond's edible-flower-goddess, Pam Vipond, in 2009 to see if she could ferment mead vinegar.

"Making vinegar is a double-fermentation process," Pam explains. Like yeast's action in breadmaking, the process needs warmth, oxygen and food. Pam adds her ten-year-old vinegar "mother," or starter, to a jug of mead. The starter contains acetobacter, the acid-based bacteria that convert alcohol into acetic acid. She introduces air through an aquarium water pump, and keeps the mead warm so the bacteria can feed on the sweet wine. The resulting vinegar is double-filtered, then pasteurized to boil off any unfermented alcohol and stabilize the vinegar before it is bottled.

Pam's vinegary is a trailer located as far from the meadery as possible, to reduce any risk of the alcohol becoming contaminated with vinegar mother. "It's every winemaker's worst nightmare, having vinegar anywhere near the wine!" she says. As a double precaution, Pam changes her clothing whenever she goes into the meadery. The process of fermenting vinegar takes six to twelve months. "You can't make it go any faster," Pam says. "And I can only take so much mead out of the meadery."

Not every mead stands up to the double process of becoming vinegar. "Melomel's fruit flavours are diffused by the process," Cherie says. "What holds up are traditional mead, and stronger flavours like blackcurrant and ginger." The team's inaugural bottles went on sale at Millarville Farmers' Market in 2010. With three one-hundred-and-fourteen-litre fermenters in use, there is a waiting list of shops anxious to carry Alberta's inaugural artisan vinegar, and Canada's first mead vinegar. As in mead-making, time is the secret ingredient.

More about vinegar

Vin agre, or "sour wine," is a preservative, a seasoning, and an esteemed palate-brightener, second only to salt in the palates of cooks. Vinegar made from frozen grapes in the style of ice wine is cherished as a digestif. In early Roman times, military troops drank watered-down vinegar while their leaders drank watered-down wine. So when the Roman soldier lifted a sponge sopping with watered-down vinegar to Jesus on the cross, it was an act of kindness.

Lola Canola Honey Bon Accord
Patty Milligan

To young Rory Milligan, "honey" is synonymous with "truck." Rory's favourite toy tanker truck is emblazoned with the word, and his mother, beekeeper Patty Milligan, says that Rory used the words interchangeably when he began to talk. It isn't surprising: in her costume as yellow-wigged Lola Canola, queen of the bees, Patty has written honey across her life in large letters.

Like wine, honey is a literal expression of the landscape, whether it's hibiscus in Jamaica or canola in Alberta. "Some people I can never make happy, because I can't give them the honey they grew up with," Patty says philosophically. "That geographic signature and its memories are what people respond to."

Lola Canola's production is determinedly small-time: from seventy hives, Patty collects ten thousand pounds annually, harvested just north-east of Edmonton in the small town of Bon Accord. Patty's commitment to staying small keeps her in her beloved house, where her mom grew up, and allows her the time to do what she loves best—educate her clientele.

Nowadays, Patty has ample to say, in stark contrast to her tongue-tied responses in 2000, her first year at a farmers' market. As a teenaged bee-keeper's apprentice, Patty learned from a bulk sales commercial beekeeper. "He was a methodical guy who taught me a lot, but when I got to the farmers' market, I couldn't answer questions about how long it will take to crystallize, or what flowers do your bees go to? My brain got going! I had worked for a beekeeper but didn't know about this stuff."

Learning about bees became an excuse to do many things—create art, talk, teach, educate. "I use the business to connect with people beyond just selling honey," she says. Often, while Patty is engrossed in conversations at the farmers' market, people will try on her yellow Lola Canola wig. (Patty has the photos as proof.)

The bees returned Patty to the farm, which gladdened her farming father's heart. "Dad said 'let's do this,' hoping to get one of his four kids into agriculture," Patty says, giggling. "So I put together a ten-year plan and got the bank loan, not pocket change but not permanent debt either."

The hives are on the farm where Patty lives, and on her dad's family farmstead at the end of the road. Like Patty, honey is place-based, and most of the honey near Bon Accord is made from the surrounding fields' canola blossoms. It's been challenging to accept that if Patty wants honey with a "sexier" profile, she'll have to live elsewhere. "Canola surrounds us. The problem is the same with any monoculture crops, or, as the writer Mark Winston says, it's like only having bananas to eat. It's just not an optimal diet." She thinks it would be way better to have a "crazy mixture"of flowers, but she has resigned herself to reality. "Bees are serial monogamists," she says, "it's easy to track where their feet have landed by what's in bloom and the colour and flavour of the honey."

Honey is graded by colour: the lighter the honey, the higher the premium on the global market. The gold-standard is clover honey. "Dandelion honey has more bounce," she says, "a little wild, not to everyone's taste." Pussy-willow blossoms lead the charge each spring; Patty took her bees' willow honey several times, but she stopped when she realized it was a vulnerable time of year and that they really needed that early honey as food.

Patty is glad to see a local shift. "Thanks to markets, Slow Food and revived interest in artisanal food, people have a growing appreciation of differences—what they won't get on the supermarket shelves, what reflects the season, the plant, the locale. It's my dream to produce wonderful, unique local varietals."

Honeybees work 'round the clock to produce honey and pollinate plants.

apple-thyme mousse and caramelized winter fruit with filo "sails"

Texture and contrast are this beautiful dessert's chief attributes. Serve the mousse chilled and the fruit warm, with room-temperature "sails." Alternatively, freeze the mousse into a semi-freddo. Ensure the mousse doesn't boil while you cook it, or the eggs will scramble. Serves 12.

mousse:
5 eggs
1 cup (250 mL) honey
¼ cup (60 mL) melted unsalted butter
1 cup (250 mL) apple cider
1–2 tsp (5–10 mL) apple cider vinegar, or to taste
2 sprigs fresh thyme
2 cups (500 mL) whipping cream
1 Tbsp (15 mL) icing sugar

fruit filling:
4 apples, peeled, cored and diced
4 pears, peeled, cored and diced
½ cup (125 mL) dried fruit, slivered or whole
(dried cherries, cranberries, apricots or pears)
3 cups (750 mL) apple cider
1 cinnamon stick
filo sails:
3 filo sheets
½ cup (125 mL) melted unsalted butter
2–4 Tbsp (30–60 mL) granulated sugar (for sprinkling)
1 tsp (5 mL) ground cinnamon (for sprinkling)
garnish:
1 tsp (5 mL) ground cinnamon (for sprinkling)
1 Tbsp (15 mL) icing sugar (for sprinkling)
12 fresh mint leaves

To make the mousse: Whisk eggs and honey in a heavy-bottomed pot. Mix in butter, cider, vinegar and thyme over medium-high heat, whisking constantly. When thickened but not yet boiling, strain through a sieve into a clean bowl. Chill, uncovered, stirring several times. Whip cream and icing sugar to firm peaks. Fold into the cold mixture.

To cook the fruit: Combine all ingredients and simmer, covered, until tender. Discard cinnamon stick. Uncover and simmer until cider is reduced to a glaze.

To make the sails: Preheat the oven to 375°F (190°C). Lay a sheet of filo on a flat surface, lightly brush edge to edge with melted butter, and sprinkle sparingly with sugar and cinnamon. Add a second sheet, buttering and sprinkling as before. Repeat with the third sheet. Slice the stacked sheets into 24 squares, six by four. Cut each square in half on the bias to form triangles. Transfer each triangle to a parchment-lined baking sheet. Bake for 3 to 5 minutes, or until golden. Cool to room temperature on the baking sheet.

To serve: Scoop cooked fruit onto 12 medium plates. Garnish each portion with mousse. Arrange 3 triangles upright on the mousse like sails. Dust with cinnamon and icing sugar. Add a mint leaf and serve immediately.

Arugula grows well in pots or beds, and is a "cut-and-come-again" green that thrives with regular harvesting.

i is for iceberg and other lettuces

The Jungle Innisfail
Leona and Blaine Staples

To Leona Staples, the world is clearly divided into two camps—those who DO, and those who don't. "The world only goes around if you help it go around," she says.

She took a calculated risk in 1996, putting the family nest egg into market gardening. "My dad told me years ago that traditional farming would not survive in this area, it would be forced out by the increasing people-population along the highway corridor. We turned that disadvantage into a strength," she says of their flourishing and well-kept farm that welcomes visitors. The Jungle is home to one of the prettiest strawberry patches in Alberta, twelve acres of summer joy.

The growing rural population of central Alberta is well-heeled, well-educated and keen on good food. Leona believes her farm will meet the needs of that population. Taking calculated risks is part of the package. She and her husband, Blaine, are constantly adding new crops to expand interest and bellies: haskap and hazelnuts developed by the University of Saskatoon, apples, sour cherries, red, white and black currants, and even a few gooseberries in the shelter offered by their house.

Blaine and Leona spent five years managing the Goldeye Centre at Nordegg, part of the Rural Education Development Association (REDA), which runs summer programs for rural youth. Leona says frankly that the networking skills she learned at Goldeye are indispensable. It's evident in her influence within the Innisfail Growers, the five-family marketing co-op with which The Jungle is affiliated.

The family grows vegetables and many greens—hydroponic lettuces that include lollo rosso, or red leaf; red oak leaf; butter lettuce; plus spinach and beets. Leona is a staunch defender of the flavour and texture of

Leona Staples, on the tractor, points out features of The Jungle, the family farm founded by her grandfather, to visitors on a *City Palate* Foodie Tootle.

iceberg lettuce. "Finding the right variety is crucial," she says, pointing to the Ithaca iceberg heads flourishing behind her. Growing lettuces hydroponically is a fluidly logical extension of their high water content, and the water eliminates the bitter finish that is common to many iceberg lettuces. It also gives The Jungle the edge in early spring. The family starts seeds in April, transplants the seedlings into their aqueous beds, and harvests before many growers have even sown seeds.

It's a simple system. A layer of Styrofoam provides bottom insulation, then a pond liner holds the water. The top floating layer of Styrofoam has holes drilled at regular intervals to provide space for the plants. The spinach outside is more subject to Mother Nature, says Leona. The Jungle is situated in the province's "hail belt," and she routinely loses three or four weeks in the spinach's life cycle after a hailstorm.

The farm's greens, berries and vegetables are sold at the farmgates of Innisfail Growers, at several farmers' markets seasonally, and at the

Innisfail Growers' year-round shop at the Calgary Farmers' Market. Leona has produced a collection of berry vinaigrettes made from raspberries, saskatoons and strawberries, with other flavours under development.

She hopes her three teenaged sons will leave home, learn a skill and then "come home" to The Jungle, the land her great-grandfather, Jacob Daniel Quantz, cleared in 1897. Any spare time goes into the family farm and whatever supports it, including stints in rural tourism initiatives within Red Deer County. And mindful of idle hands and dwindling knowledge, Leona, the busy farmer with no spare time to speak of, offers classes in the farm wife's forgotten skills, pie-making and preserving, in her efforts to keep the world spinning.

iceberg, arugula and orange saladio with pink pickled onions and pine nuts

A warm vinaigrette based around orange juice and nut oil enhances iceberg's sweet nature, with arugula and pickled red onions for balance. Serve this salad alone or as a garnish for pasta, pork roast or grilled lamb. Pickled red onions tame oniony bad breath and add brilliant pink to any dish. Store leftover pickled onions in the fridge for up to ten days for optimal crunch; longer storage makes a softer pickle. Use them on anything that needs the crunchy presence of mildly sweet pickled onion. Serves 2–4.

dressing:
2 bacon slices, finely slivered
¼ cup (60 mL) orange juice
2 Tbsp (30 mL) sherry vinegar or balsamic vinegar
1 Tbsp (15 mL) mustard
2 Tbsp (30 mL) walnut oil
2 Tbsp (30 mL) olive oil or cold-pressed organic canola oil
kosher salt and cayenne to taste
pink pickled onions:
1 finely sliced red onion
1 cup (250 mL) white wine vinegar or rice vinegar
1 tsp (5 mL) sugar
kosher salt and hot chile flakes to taste
saladio:
1 cup (250 mL) iceberg lettuce, leaves torn or sliced
1 cup (250 mL) arugula leaves
1 orange, sliced or segmented
¼ cup (60 mL) toasted pine nuts
slices of your favourite cheese (optional)

In a frying pan, sauté bacon until crisp, then remove bacon from the pan. Include the bacon fat in the dressing, or discard it if you prefer. Off the heat, mix together vinaigrette ingredients, adding the bacon. Set aside.

Place onion slices in a colander in the sink. Pour several cups of boiling water over them and discard the water. Transfer hot onions to a glass or non-reactive bowl. Cover slices immediately with vinegar, sugar, salt and hot chile flakes to taste. Mix well, cover and chill.

Toss lettuce, arugula, orange and pine nuts. Toss well and arrange on plates beside an optional slice or wedge of cheese. Top with pickled red onion slices. Serve.

j is for jalapeno and other chile peppers

Broxburn Vegetables & Café Lethbridge
Paul and Hilda de Jonge

It's a balmy autumn evening in 2002 when I first visit Paul de Jonge's farm. I head into the field and kneel in the warm soil, picking and eating strawberries with my chef friend Scott Pohorelic in the u-pick patch. The red berries on their runners are illuminated by late evening sun, that long, slanted, almost surreal light of the Prairies.

In the greenhouses, the light is diffused and detached into separate rays that angle around the long rows of jalapeno and bell pepper plants. Broxburn Vegetables & Café co-owner Paul de Jonge says the greenhouses cover over three acres, nearly one hundred and fifty thousand square feet, or the ice surface of eight NHL hockey arenas. En route through the massive buildings, he pulls sweet and spicy peppers from the plants for me to taste, then we cross the veranda and sit down in the café. After a bowl of roasted red pepper soup and a slice of strawberry pie, I agree when Paul says, "Better than ordinary." And although he's talking about his café's soup and pie, the term applies to Broxburn produce too.

Paul's farm family left the Netherlands when he was seventeen. "My father was successful, but it was absolutely regimented there. At age six I was in the onion fields. To me, still, the smell of a cut onion is perfume." At age twenty-eight, Paul and his wife, Hilda, bought the farm northeast of Lethbridge. Hilda planted a u-pick strawberry market garden with a neighbour. "Strawberries, because I like eating them, and if you plant spuds you just can't eat them all if you don't sell 'em all," Paul rationalized. Hilda did all the work, he recounts, but by 1999, the market garden was so successful that Paul gave up his off-farm accountant's job. The first greenhouse went up a year later.

The couple learned early at the local farmers' market that specializing

Jalapeno peppers grow in greenhouses at Broxburn Vegetables & Café in Lethbridge.

was the route to take. They filled the greenhouse with cucumbers and peppers, and grew onions, cauliflower and its cousins, cabbage and broccoli, in the fields. When u-pick customers insistently asked to buy peppers, they turned an old barn into a handsome store framed by a veranda, with a daytime café to use up the "number two" produce. "It takes a lot of roasted peppers to make soup, and we only sell the number ones," Paul says.

In 2004, Rudy Knitel, whose Italian import business, Galimax Trading Inc., was struggling, asked Giuseppe DiGennaro, a high-profile Italian-born chef in Calgary, to visit Broxburn. Giuseppe loved the red pepper soup, and inquired about buying produce. Before he drove away, a deal had been struck, and Rudy began to broker and deliver Broxburn's produce to Calgary chefs. Paul added tomatoes a year later, and in 2006, Rudy started bringing busloads of chefs to visit Broxburn and several nearby farms.

Paul is justifiably proud of his relationship with Rudy. "We pick today, and tomorrow our produce is on the chefs' counters. It's a short cycle with big flavour," he says. "A big supplier just can't do that." The alternative, supplying Medicine Hat's Red Hat Co-op, would be easier than loading the trucks at night for their twice-weekly run, but Paul says his restaurant customers are significant clientele. "Of course we get more money, but we guarantee the quality. I never get any produce back from those trips."

Nearly eighty Albertan restaurants, from Calgary to Lake Louise to Banff, proudly use Broxburn produce—cucumbers, tomatoes, brassicas, and Paul's magnificent jalapenos and sweet bell peppers. Paul and Hilda are blessed with a longer growing season than Calgary due to lower elevation, and use integrated pest management (IPM), a system that controls pest insects with beneficial ones.

"I am not a chef, I am a farmer, a jack of all trades. But I don't let an opportunity go by," says Paul, who sells his farm's produce at four farmers' markets in Calgary and Lethbridge. "Markets are where I see the most potential, where customers can ask for and get honest answers. Part of the reward is that people come back week after week. It would be great to have all those people come to the farm, but it's just too long a drive. A restaurant is a business, where chefs have to make something from those peppers *and* make money. But at the market, chefs buy peppers just to feed their family." Like the light on his strawberry field, that intrinsic reward keeps Paul de Jonge going.

What may look random or untidy is actually carefully considered spacing of heavy horse implements, left lying on the ground with sufficient room for two-thousand-pound horses to be positioned for simple hookup.

Oxyoke Farms Linden
Robby and Phyllis Fyn

"People ask me, how do you grow peppers, sweet or hot, in Alberta? I don't understand why people have a problem," Robby Fyn says. "I do what my dad did, he did what my grandfather did, and *he* did what my great-grandfather did. People make it too complicated. It isn't. My great great-grandfather wrote some of his farm practices in his diary, and now farmers are doing them again. It's nice they're being discovered, but don't tell me they're new."

Robby Fyn, raised in Kitchener-Waterloo to a German Baptist family that used horses and buggies for transport, planted his first garden plot when he was eight. His mother told him he could have eight rows. Robby calculated what crop was most valuable at the market, planted head lettuce and took his first greens to market while he was still in three-quarter-length pants.

Robby's in his early sixties now, with a long silver ponytail. That's just about the only thing that's changed, he tells me, laughing, other than noticing a smaller fork-load when he uses his pitchfork in the barn. Robby, his wife, Phyllis, and their daughter Barbara tend a CSA garden on their quarter section, bisected by Kneehill Creek, eighty kilometres northeast of Calgary. The farmstead sits on a ridge, land falling steeply northward at its back, with a gradual sloping south side. There, where it is always dry, they planted the beds. It's a busy place, with adult kids and their seven grandkids checking in pretty much daily.

"Our older way of doing things is still good. We are traditional, natural farmers. We don't use any commercial sprays or fertilizers. But we don't use 'organic.' The word is getting so mutilated, it's not going to have any validity down the road. The problem is when companies like Monsanto start to get involved, it's like a fox has been hired to keep an eye on the henhouse.

"You have this land for just a short while," Robby says. "It's your responsibility, but that gives you no ownership. It belongs to your grandchildren, and you are just a steward. Treat it with dignity and respect, and you'll be surprised at how friendly that ground will be."

Part of how Robby respects the land is to tread lightly on it, using a plough hitched to oxen. "There's a different rhythm behind a plough, the snapping of the roots when the shear cuts them off. The poet Robert Burns said it well, about a mouse scurrying in the disturbed furrow ahead of the plough. You'd never hear that on a tractor."

There are other ways to show respect, he says. Treat your Berkshire sows to a straw-filled box stall when they farrow. Feed your chickens grain in the stock trailer to keep them calm, a last rite of affection before they go to the slaughterhouse. There's no folksy answer to what keeps him in the farming game, he says. "I get up in the morning and feed the livestock. I enjoy it. I eat good food. I like to think we produce good food."

Oh, and about the peppers: Robby's last word on the subject is this: "I don't know why people don't garden. I could be harsh and say they are lazy. Or arrogant, and say they think they are better than that. But what I really believe is that they don't realize they can do it. All you have to do is create a small microclimate. Put up a wind fence, and use biodegradable black plastic to hold in the heat and deter the weeds. And as my mom said, you got lemons, make lemonade." Just what lemonade even Robby Fyn's mom would make of quackgrass and other weeds isn't exactly clear, but as her son says, the very presence of quackgrass indicates a healthy soil. And the ecstasy of soil healthy enough to grow jalapenos and other vegetables is what farmers long for.

Gull Valley Greenhouse Gull Lake
Phil Tiemstra

Phil Tiemstra knows who his friends are. His pesticide-free greenhouses are home to beautiful vegetables, and to insects that provide biological controls for what might otherwise become a problematic population of the bad sort of bugs. So Phil's arm's-length friends include ladybugs, parasitic wasps, mites, midges and their larvae—all used for aphid, spider mite and whitefly control—and bumblebees for pollination.

Thai bird peppers are grown in biologically controlled greenhouses, where the parasitic wasp *Aphidius colemani* controls aphids by stinging and mummifying them.

"We try to introduce small numbers of insects," he says. "The key is to be one step ahead of the problem."

Phil's greenhouses, over the hill and down the valley close to Gull Lake, in central Alberta, are filled with hot and sweet peppers, heirloom tomatoes, eggplant, green beans and flat roman beans. A group of hard-working labourers from Mexico, Thailand and Chile varies in number, from seven in January to sixteen in summer and fall. Phil estimates he spends 20 per cent of his time on paperwork to facilitate their easy travel to and from their home countries each year. "The same people each year," he says, pleased. "They come back in April when the picking starts in earnest. One couple wants to stay. I'm looking into it under the Alberta Nominee Program. They have some solid supervisory skills."

His Thai staff planted some tiny incendiary Thai bird peppers to go with the jalapenos and Scotch bonnets they already grow. "No, it's not for me, that hot food," Phil says. "But my son-in-law, Scott Epple, who takes the vegetables to the farmers' markets, he likes it."

The peppers in particular are prone to aphids. The most spectacular biological control that Phil uses is the parasitic wasp *Aphidius colemani*. It stings the aphids, then mummifies them. What emerge from the bronze-coloured mummies are not aphids, but more wasps, veritable cuckoos in the nest.

Other bugs do other jobs. "We use *Aphidoletes* to consume the aphid honey dew," he explains. "The adults are voracious, and their larvae have a big appetite too. For the whiteflies that pester the tomatoes, we get *Encarsia formosa*." This predator wasp, which lays its eggs on whitefly scale, was commonly used in greenhouses prior to the emergence of chemical pesticides.

Phil was a poultry farmer with no plans to be a greenhouse grower, but in 1992, he toured some greenhouses north of Edmonton. The idea and the challenge excited him, so he found a mentor. The new project started small, with five other associates who formed a marketing co-op known as Pik'n Pak Produce, and six greenhouse buildings that covered eighteen thousand square feet in total. Phil chose heirloom varieties of vegetables that he thought would have commercial appeal. "They're no more work than commercial varieties," he says. "We've tried a bunch. Some worked, others not all." Green Zebrino cherry tomatoes are zesty, with a citrusy bite and a purplish interior, a variety that Phil is partial to cooking and barbecuing. "Orange cherry tomatoes, called SunSugar, they have a real sweet and sharp contrast, two things going on at once."

A decade later, the buildings' footprint has multiplied by a factor of nearly ten, covering almost three acres. In the early years, they sold almost everything wholesale. But then greenhouse businesses took off during the late 1990s in BC and southern Ontario. "We had to lower our prices to compete with increased competition," Phil says. That's when Scott started selling at the farmers' markets, first in Edmonton, then in Calgary.

"We're very happy about the markets. We wouldn't be in business without them."

They wouldn't be in business without their small insect friends either. "We've pushed hard for pesticide-free plants," Phil says. "The insects have been a steep learning curve."

Sweet or spicy peppers serve as a delicious holder for nuts, seeds, roasted cauliflower and cheese.

desert stuffed peppers with pepitas and gouda

This Moroccan-influenced approach to stuffed peppers is light and delicious. Using seeds and nuts adds a little protein hit for hot days when meat or chicken is simply too heavy. Use hot, bullet-nosed jalapenos or mild, elongated Anaheim, banana or pasilla peppers as the casing if you are so inclined. For a dish with no residual heat, choose thin-skinned Shepherd or sweet Hungarian peppers. Serves 3–8 as a starter or appetizer (using small hot or mild peppers), or 3 as a light lunch (using medium-sized mild peppers).

3 mild bell peppers or 8 hot peppers, halved lengthwise and seeded
1 Tbsp (15 mL) olive oil or cold-pressed organic canola oil
    (for brushing onto pepper skins)
2 Tbsp (30 mL) olive oil or cold-pressed organic canola oil, for the pan
½ onion, minced
1 garlic clove, minced
1 tsp (5 mL) minced jalapeno or other hot pepper (optional)
½ cup (125 mL) finely chopped roasted cauliflower (see Cook's note)
½ tsp (2.5 mL) roasted ground cumin
¼ tsp (1 mL) dried oregano
⅛ tsp (0.5 mL) ground cinnamon
1 cup (250 mL) arugula, coarsely chopped
2 Tbsp (30 mL) toasted pine nuts
2 Tbsp (30 mL) raw sunflower seeds
2 Tbsp (30 mL) chopped parsley or cilantro
juice and zest of 1 lime
kosher salt and freshly cracked pepper to taste
1 cup (250 mL) grated Gouda cheese, divided

Preheat the broiler. Place peppers, skin side up, on a baking sheet and lightly brush with oil. Broil until skins are wrinkled and beginning to char. Remove from heat. Flip the peppers so they are skin side down on the baking sheet. Set aside.

Heat oil in a sauté pan. Add onion, garlic, minced hot pepper, cauliflower, cumin, oregano and cinnamon. Cook until tender, about 5 minutes. Stir in arugula and wilt it. Add pine nuts, sunflower seeds, parsley or cilantro, lime juice and zest, and season to taste with salt and pepper. Mix well, then stir in half the cheese. Divide the mixture among the halved peppers. Top with the remaining cheese and broil until melted. Serve hot or cool.

Cook's note: Divide the cauliflower head into florets, discarding the core. Toss the florets lightly in olive oil or cold-pressed organic canola oil and spread in a single layer on a parchment-lined baking sheet. Roast at 375°F (175°C) for 30 to 45 minutes, until golden brown, stirring once or twice during the cooking process.

Kale's hardy nature adds dense texture and welcome colour to fall gardens and menus.

k is for kale

Thompson Small Farm and Bergen Farm Sundre
Jonathan Wright and Andrea Thompson

> "Whilst watering the garden the other day, a peaceful and protracted process that leaves lots of room for contemplation, I looked down at a kale bed and suddenly felt I was in a light plane over the African Veldt. The kale were like little acacia trees over the tawny scrub."
> —from the Thompson Small Farm website

Children sing about the farmer in the dell. But until I first visited Thompson Small Farm in the winter of 2008, I believed such things belonged in nursery rhymes. How wonderful to be proven wrong in such a beautiful way! This small farm is just that, a sinkhole for heat, its sloping aspect facing south. In the following months, I drive to the farm, not to collect my greens—CSA farmers Jonathan Wright and Andrea Thompson deliver them weekly to their city-dweller clients—but because I like the chance to lean against the Clydesdale mares and help adjust their harnesses, to imagine the feel of the leather reins in my hands, to consider learning to drive them and see the land open beneath the metal blade of the plow. My grandfather cut open the prairie behind a team of heavy horses. There are few farmers left who know how to do so.

In late autumn, Jon and Andrea's website offers draft horse driving lessons as well as animal tracking services. By then, the harvest is in, and the couple's extended network, the subscribers who receive a weekly share of the crop, have all been fed the final week's harvest.

In the fall of 2008, after their first year as CSA farmers, the couple wavered and nearly gave up, but decided to keep on after meeting Robby Fyn, a farmer who still uses draft animals in his fields. For three years, a family subscription

Despite their massive size, heavy horses like Clydesdales Raven and Gwenyth tread more lightly on the ground than a tractor, and allow Jonathan Wright to till the fields at a slower pace of life.

provides me and my youngest son with eighteen weeks' worth of abundant produce raised without chemicals or hydrocarbons, all planted according to the biodynamic calendar. They sell twenty-four subscriptions in a season, which will never make them rich, but the funds raised take them through to the following spring. No hydrocarbons means that draft horses and water buffalo provide the pulling power normally associated with tractors.

The original twenty-acre farm, astride gentle hills east of Beiseker, is home to several greenhouses, three acres of newly broken gardens in a sheltered, south-facing dell, and a newly planted orchard on a north-facing slope. A small herd of yaks graze with four Clydesdale mares. A pair of water buffalo, purchased from Fairburn Farm on Vancouver Island, walk past with horned heads held high, headed for pasture.

In the winter of 2011, the couple pulled up stakes and moved, to a new spot near Sundre, seeking access to timber, a larger piece of land and closer access to their Calgary clients.

Draft animals dictate a slower pace, Jon observes, and they tread on the earth more lightly than machines. They eat hay and grass, and don't rely on depleting fossil fuels. "Horses are better company than a tractor. For us, this is a move toward making better choices that are sustainable," he tells me as we tour the farm on a chilly day that shows little sign of spring thaw.

Old disc harrowers, plows and mowers topped with unforgiving metal seats lie on the verge of the driveway, looking as if they've been dropped carelessly. In reality, each is carefully placed, with sufficient space for hitching to one or two mares weighing nearly two thousand pounds each.

Using draft animals means that the couple had to find not only a mentor to train them in their care and use, but equipment too. Jon went to old farmers, collectors and estate sales to locate cultivators, carts and harnesses for his large mares.

My son and I eat chard for weeks on end. As the weather turns, the kale starts arriving, a denser, rougher leaf structure, a more regal purple than chard's democratic rainbow, and slower to cook too. Overabundance and repetition is how I read, how I listen to music, and for three years, with the help of a CSA gardener, how I cook and eat.

"I never wanted a career, I wanted a life," Jon tells me. "The farm lifestyle is three-dimensional for me, a work in progress. Paint the barn. Finish breaking the filly, fence the orchard. Plant the berries, turn the sod. Plant the seeds."

Sparrow's Nest Organics Opal
Graham Sparrow

Graham Sparrow was a twenty-six-year-old Canadian city boy working in rural Ireland when his viewpoint was altered forever. "I saw the most amazing families, home-schooling and baking bread, chickens coming into the kitchen," he says in wonderment. "And beautiful, dirty, red-cheeked kids playing. I thought, 'this has meaning to me. This is not how I grew up at all.' From that, I realized that everything related to becoming a farmer and growing food."

Now in his early forties, Graham recalls the struggle he underwent when he returned to his parents' Edmonton home in 1995 after backpacking through Europe. The economics grad knew he wanted to leave city life, and found summer work tending market gardens, first in Nelson, then in Kamloops. "Kamloops was my first real farm experience," Graham says. "I learned so much."

The hook was set. Graham found himself attending a sustainable-farming conference in California, then working on a ten-acre biodynamic tree orchard and garden in Monterey. "It was pure synchronicity," he says. "I learned all about drip irrigation, and I helped to start a small CSA." A job in Kaslo, BC, was followed by a fallow patch, during which Graham lived with his brother Ron, north of Edmonton.

"Ron kept encouraging me to start my own farm." Graham says. He finally settled on seventy acres, with a house perched one hundred feet above the Redwater River valley, north of Edmonton. Graham did a hand-shake deal with the owner and moved in.

After three years transitioning to organic, Graham let go of his day job driving a front-end loader at his neighbour's gravel pit and leapt completely into market gardening. He took control of his rocks and quackgrass, bought a small tractor and cultivator, and opened his CSA in 2001 with thirteen shares. In 2011, he was up to eighty-five shares, mostly Edmontonians. "Redwater residents are in the oil industry or they're ranchers. People have big gardens there, and don't want to pay for organics. The urban population is interested in supporting organics."

The five-acre garden rotates yearly to a different spot in a twenty-acre patch, yielding perennials, herbs, root crops, and hot-weather Mediterranean vegetables like eggplant, peppers, tomatoes and basil that Graham seeds under reusable fabric wind tunnels on the fields, like long caterpillars on round metal staves. "Kale flourishes here," he says. "My fave is Light Russian, it's tender, and mild, not like dinosaur (Tuscan) kale, so tough it's still chewy after fifteen minutes of steaming."

Gardening is the act of optimists. "Everything is complicated. I'm waiting for that year when everything goes smoothly. This is my eleventh year with market gardens, with three years of serious drought at the beginning. There's always something." Despite that, Graham says that sustainable farming feels like a right-livelihood, Buddhist thing, good for souls and soil and people. "It's a real alternative to agribiz. But it all has to change. Look at the small growers who are barely keeping it together. Everything seems to be supporting the wrong thing; it's the unsustainable fossil fuel system that will be falling apart and that we should be preparing for. I'm not looking to be subsidized, but I'm competing against huge systems. What we need are small niches of people concerned with what they are eating. Once they come to the farm, they don't go back."

chicken ballotine stuffed with kale, mushrooms and sage

This stuffed and roasted dish is roundly flavoured and earthy. It reheats very well, so make extra. Use the stuffing in cannelloni, lasagna and ravioli. Substitute other greens like chard, arugula or spinach for the kale if you wish, altering the cooking time to suit the texture of the greens. Serves 6–8.

stuffing:
6–8 chicken thighs, bone in and skin on
3 Tbsp (45 mL) olive oil or cold-pressed organic canola oil
1 leek, sliced
1 large carrot, coarsely grated
1 medium parsnip, coarsely grated
4 garlic cloves, minced
½ cup (125 mL) sliced crimini, porcini or chanterelle mushrooms
½ cup (125 mL) sliced field mushrooms
3–6 kale leaves, depending on size, finely shredded
1 Tbsp (15 mL) finely minced fresh chives
1 Tbsp (15 mL) minced fresh thyme
1 tsp (5 mL) minced fresh sage
kosher salt and pepper to taste
olive oil for drizzling

sauce:
2 Tbsp (30 mL) olive oil or cold-pressed organic canola oil
roasted chicken bones from thighs (plus extra roasted bones if available)
1 carrot, minced
1 celery stalk, minced
1 onion, minced
1 Tbsp (15 mL) tomato paste
1 tsp (5 mL) flour
¾ cup (200 mL) red wine
1 tsp (5 mL) minced fresh sage
3 cups (750 mL) chicken stock
2 Tbsp (30 mL) unsalted butter
½ cup (125 mL) sliced field mushrooms, crimini, porcini or chanterelles
3 oz (90 mL) brandy

Preheat the oven to 400°F (200°C). Remove bones from chicken thighs. Set bones on a heat-proof baking sheet. Roast bones, uncovered, until brown. Set aside.

To make the stuffing, heat half the oil, then add leek, carrot, parsnip and garlic. Sauté until tender and wilted. Add mushrooms and sauté for several minutes, until tender. Add kale. Cover and steam until tender, checking every few minutes. Add half the herbs, salt and pepper to taste. Set aside.

When stuffing is cooled, lay chicken, skin side down, on a clean counter or large tray. Season with salt and pepper. Spoon the stuffing lengthwise down the middle of the meat. Fold or roll up and place on parchment-lined tray, skin side up, with the seam on the bottom. Drizzle with remaining oil and season to taste with remaining herbs, salt and pepper. Roast in a 400°F (200°C) oven for 20 minutes or until juices run clear. Set aside, loosely covered.

While chicken roasts, make the sauce. Heat oil in a saucepan, then add bones, carrot, celery and onion. Sauté until brown. Stir in tomato paste and cook for 1 minute. Add flour to coat all ingredients, mix well and cook briefly until sandy in texture. Deglaze with red wine, stirring well to dislodge any browned bits. Add sage and stock. Cook over medium-high heat until the sauce is reduced by three-quarters. Strain through a fine strainer into a pot and keep warm. Heat butter in a small sauté pan, then add mushrooms and sauté until tender. Add brandy to the pan and flambé until alcohol is burned off. Add to reserved sauce.

To serve, after chicken ballotines have rested for 5 minutes, slice each one in half on the bias. Tip any juices from the roasting pan into the sauce. Set one piece of chicken on each plate and spoon sauce around but not over the meat. Serve immediately.

l is for lamb

Driview Farms Fort Macleod
Gerrit and Janet van Hierden

In his mid-fifties, Gerrit van Hierden is blessed with the easygoing nature that animals and children respond to best. He's spent thirty-five years raising four daughters and countless sheep and practising patience. But a perfectionist's steel is concealed behind the mellow demeanour. Gerrit's just come in from putting up a fence. It's windy in the south country, and the eighteen-foot fence will protect the new house that he and his son-in-law Bert built so the close-knit family could be closer to Gerrit's wife Janet's parents. The house sits a mile from the Belly River, a bit higher, so they can see around the country. "We did a lot of the work ourselves. Once you start putting up rafters, you want to make sure everything is done right," Gerrit says.

Sheep were in the frame from the start. "When I married Janet, a horse and a sheep came along with her. A dowry, I guess," he says, laughing. "Later we got lambs for our girls to feed so they could learn to take care of animals." The flock grew haphazardly as Gerrit bought ewes here and there. After a livestock auction added two hundred animals to the flock in one fell swoop, Gerrit realized that what had been a hobby had grown into something else. "Janet was running out of names," he says wryly. "We decided then that we needed to run it as a business or cut back." The flock now numbers about six hundred and fifty animals, plus three large guard dogs.

"Sheep can teach you a lot of things if you let them," Gerrit says, sounding rueful. "They're intelligent in their own way, but they do things from curiosity. If you are fighting with them, it's up to you to change how you deal with them. Animals are like people. You have to get them to do what you want them to without them knowing it."

That observation proved useful in 2001, when Gerrit started selling lamb at the Millarville farmers' market. "I thought, 'there are people who will appreciate the way we raise our lamb.'" It wasn't long before Gerrit was making weekly trips to Calgary, making deliveries to a select list of restaurants, and occasionally being met en route by private clients looking to stock their freezers. "I enjoy the one-to-one with my customers. They appreciate what we are doing to raise a good lamb. And that gives me a lot of satisfaction."

Restaurant chefs like Bob Matthews at La Chaumière in Calgary call Driview lamb the best in Alberta. "I've discussed why that is with our chefs. I thought it was the hay and the grain we raise ourselves, but Ken Titcomb [of The Ranchmen's Club in Calgary] thinks it's the water. We're on a well, our animals don't drink from a dugout, so our water is super clean. It's hard water, a high mineral content." He shrugs and says, "Happy sheep are healthy sheep. Healthy sheep make you more money."

Money is always an issue. "In all stages of agriculture, corporations are taking over more and more. The government keeps making new rules that corporations can afford to comply with; while on a family farm, we don't have the resources, the people, the expertise, and we can't necessarily hire someone to do the paperwork." Gerrit's pragmatism resurfaces. "What I make at farmers' markets, I lose on the farm. But I enjoy the rural life; it has a peace and quiet you can't find anywhere else. But I can't retire on it. My farm is my retirement, some of it will have to be sold someday. We can leave the kids the rest."

Cakadu Heritage Lamb Innisfail
Denis and Linda Jabs

Halfway through a day-long tour of central Albertan farms, forty-eight people on *City Palate*'s annual "Foodie Tootle" bus need to stretch their legs. As they shuffle off the bus, the humans are greeted by a flock of frisky, long-legged creatures beyond the wooden fence. As the flock races from one end of the field to the other, the effect is like water receding and advancing. Any hope of counting heads or tails is immediately abandoned.

Our hosts, Linda and Denis Jabs, surprise us when they explain the sprightly beasts are not goats, but Barbados blackbelly sheep that originated in the West Indies in 1627. "We first saw them tethered on a lawn in Barbados in 1992, a year before we bought the farm, and we thought they were goats too," Linda says to the group. "Four months later, Denis came home from an auction with a ram and six ewes. 'What are we going to do with those?' I asked him. We'd planned to buy bison, you see. He said, 'I don't know but I liked the curl of the ram's horn.'" At present, the five-hundred-animal flock includes six handsome rams. Their horns are magnificent, backswept and downward-curving as if slicked by the wind.

Later that evening, the bus tourists enjoy mild roasted meat that is not overtly lamb-like. "The Barbados blackbelly is one of the 'hair' breeds of sheep," Linda explains. The animals are covered by coarse hair, not wool, and they don't have lanolin, the waxy substance that keeps woolly sheep waterproof and imbues their meat with its musk.

Linda and Denis joined Rare Breeds Canada to preserve what has become a rare breed. The sheep are easy to maintain, and highly prolific, Linda adds. "We don't have to interfere too often. The ewes mostly throw twins, and they bond immediately to their lambs."

Each evening at sundown, about two hundred lambs run from one end of the field to the other for ten or fifteen minutes. Then they all lie down. "Are they happy to be alive or getting ready to sleep?" Linda wonders. "We don't know, but it's our favourite thing to watch; it makes us laugh." Two Italian-bred Maremma sheepdogs protect the free-ranging flock from coyotes.

Denis and Linda grew up on farms. Denis retired in 2011 as a correctional officer at Bowden Institution, and Linda is a project manager and facilitator with Clean Air Strategic Alliance, an organization of businesses, government and individuals committed to managing and improving air quality in Alberta. "How do we help the whole support itself and be sustainable? Joining Slow Food put us in contact with people with the same beliefs. We aim to keep the land as productive as we can, sustainably."

A wildlife corridor at the foot of their quarter runs from the Red Deer River to the nearby Medicine River. Frequent inquiries remind Linda that they live in a province based on natural resource exploitation. "We make every effort to keep power and oil corporations off our land to preserve it. That's the toughest thing we've had to deal with—protecting water and land they want from seismic crews, power lines, methane drilling, oil and natural gas extraction," Linda says. "They can directional drill on adjacent land so they know what's here, and they don't need permission to take it.

Denis Jabs likes the "wild" nature of his Barbados blackbelly sheep, who look surprisingly goat-like. PHOTO: KAREN ANDERSON

The best we can do is ask for a lot of environmental requirements before they come on our land and after they leave. The rest, we don't believe in it and won't have any part of it if we don't have to." She and Denis far prefer to watch the sheep playing at bedtime.

Ewe-Nique Farms Champion
Bert and Caroline Vande Bruinhorst

Bert and Caroline Vande Bruinhorst's eleven children, toddlers to twenty-somethings, occupy the central role in the family's decision-making processes. "It's not a big money business," Caroline says of their sheep farm. "We're doing it to see our kids grow up with animals and work with them. And so they know how great it is to live in the country. You can't let little kids out with bulls, but sheep are pretty safe."

Bert and Caroline are offspring of Dutch immigrants to the Coaldale area of Alberta. "We met as teens, at the Netherlands Reform Church," Caroline recalls. Church has a major influence in their lives: Caroline home-schools the school-age children using CDs and distance-learning curricula from A Beka Academy in Florida. "Bible teachings and religion are integrated in every aspect of our lives. We hope it influences our children to make the right choices. Home-schooling teaches self-motivation and how to work together."

"Having a large family is a pile of work," Caroline adds. "But you get to a point where you're busy anyway, so what's a little extra?"

The kids do a lot; as a way of keeping them involved, the couple's two eldest twentysomething daughters own the sheep. Their daughters' diligent work and boundless energy are part of the family farm's success. "They want to expand to one thousand ewes; they just have that much energy. When they were younger, they got all their schoolwork done by one, to be out all afternoon with Bert. They are better at farming than he will ever be. We see a team effort in the future, absolutely."

The family moved to their current section of land near Champion, east of Nanton, in 2007. Leaving irrigation country near Picture Butte to dry-land country made a big difference. "It means not having to get out of bed at 5:00 AM to move the irrigation wheel line at the pivot, then twice again before nine o'clock." Moving the animals from one grazing pasture to the next is child's play. "Just fill a pail with barley and they'll follow you," Caroline says, chuckling. She uses milk from the family's pair of Holstein cows to make cheese, yogourt, kefir and butter. The younger children start on schoolwork before breakfast, a leisurely meal with time for morning devotions and the allocation of daily responsibilities.

The family works in the on-farm meat shop two days a week. "It's an assembly line in there: Bert cuts the lamb in quarters; our older children cut and trim the legs and the shoulders; and two do the packaging, vacuum-sealing, and into the cooler." Weekends are family time, and Sundays are strictly nothing but basic necessities, including two trips to Lethbridge for church.

Ewe-Nique Farms has formed a co-operative delivery and marketing alliance with two neighbouring farms, Broek Pork Acres and Noble Duck, the latter owned by Caroline's brother Gerwin Vandeuveren. Every Thursday, the jointly owned refrigerated truck makes its way to restaurant clients in Calgary and Lethbridge. On the return trip, the empty truck stops at the abattoir in High River to collect the week's twenty lamb carcasses to hang in the farm's walk-in cooler for five days.

For Caroline, the effort has its own reward. "We see the fruit of our labours pretty clearly—our kids, all those sisters and brothers, are best friends. We have so much fun with our kids, and we're so impressed our kids like to be home with us. We are a whole family. That is something we have never regretted."

Barbados blackbelly sheep are one of several species of "hair" sheep, without lanolin or wool, the absence of which result in mild meat, the "veal" of lamb.

rogan josh

My version of this classic Kashmiri and Punjabi lamb dish cooks slowly in the oven for several hours. For a quicker-cooking dish, use lamb chops or loin. Adapt to other meats by substituting bison hump, pork shoulder, chicken thighs or beef chuck. Serve with chutney, basmati rice and heated flatbread (tortillas or naan). Serves 6.

marinade & meat:
1–2 bird chiles, chopped
8 garlic cloves, minced
2 Tbsp (30 mL) grated ginger root
juice of 1 lemon
3 cups (750 mL) plain yogourt
2 lb (900 g) lamb shoulder, sliced into 2 in (5 cm) cubes
1 Tbsp (15 mL) kosher salt

braising sauce:
2 Tbsp (30 mL) olive oil or cold-pressed organic canola oil
1 large onion, minced
6 garlic cloves, minced
1 Tbsp (15 mL) grated ginger root
1 tsp (5 mL) roasted and ground cumin seed
1 tsp (5 mL) roasted and ground coriander seed
½ tsp (2.5 mL) fennel seed, roasted and cracked
½ tsp (2.5 mL) chili powder
2 Tbsp (30 mL) garam masala or curry powder
2 Tbsp (30 mL) tomato paste
kosher salt to taste
juice of 1 lemon
½ tsp (2.5 mL) finely grated lemon zest
2 Tbsp (30 mL) minced cilantro
2 Tbsp (30 mL) minced spearmint

Combine the marinade ingredients and meat in a large bowl and let stand, uncovered, 15 to 30 minutes, or as time allows. Preheat the oven to 300°F (145°C). To make the sauce, heat the oil in a large oven-proof braising pan, then add onion, garlic and ginger. Cook over medium-high heat until brown and fragrant, about 10 minutes. Add meat, marinade, spices and tomato paste. Mix well, bring to a boil and cover snugly. Transfer to the oven and cook for 3 hours, or until tender. Skim off any fat. Balance with salt and lemon juice. Garnish with lemon zest and herbs just before serving.

In Europe, designations such as the Italian DOP (*Denominazione di Origine Protetta*, or Protected Designation of Origin) and French *Appellation d'Origine Protégée* (AOP) protect artisanal methods of food production. These pecorinos made in Canada by Rhonda Zuk Headon have no such system of designation.

m is for "milk's immortal leap": cheese

(with thanks to Clifton Fadiman)

The Cheesiry Kitscoty
Rhonda Zuk Headon

Rhonda Zuk intended her trip to Europe to be life-changing. The Edmontonian had promised herself that if she wasn't married or engaged by age thirty, she'd go looking for adventure. What Rhonda didn't know was that an off-the-cuff visit to a cheese farm in Tuscany would unexpectedly change the template of her life. In 2007, only a few Albertan cheesemakers were practising the ancient equation of time plus milk equals cheese. Rhonda wasn't among them.

Before she flew overseas, Rhonda re-met Brian Headon at Vegreville's annual *pysanka* (Ukrainian Easter egg) festival, watching the colourful swirls of the *kolomayka* (circle dance). They had first met at a hockey game in 2004. During the second encounter, the pair exchanged phone numbers and subsequently began to date.

Rhonda's first few weeks in Italy were spent in Tuscany with her mother, Sharon. While there, their *agriturismo* hosts directed them to a local restaurant to eat a local specialty soup, *ribollita*, and to a cheese farm in Pienza, for a tour and tasting of Tuscan pecorino. When Brian joined Rhonda in October, she left Amalfi, Sicily and Rome behind to take him back to the Tuscan dairy farm.

After the two returned to Canada, the cheese farm preoccupied Rhonda. She returned for four months during 2008, taking language lessons in Florence, making ribollita, and volunteering for six weeks at the farm under the keen supervision of itinerant Austrian cheesemaker Don Ronaldo.

"I spent my time washing and turning cheeses in the aging room, and cleaning equipment. Each evening we walked to the nearby winery to drink wine on the terrace, talking about cheese," Rhonda recalls. She had

Rhonda Zuk Headon learned to make pecorino in Italy, then came home to Alberta and set up a cheese-making facility on the family farm near Lloydminster.

neither equipment nor animals, but she knew she wanted a cheesery. In 2009, newly married, Rhonda began making plans. Cheese-making was to be her contribution to the farm where her husband's family lived. The family had milked dairy cows in an earlier era. "They were fairly positive," she says, "but they were concerned about me, a rookie, running a dairy *and* making cheese while Brian ran the beef cattle. Each is a full-time job."

The unspoken question loomed. How would a local cheese be received in small-town Alberta?

The old dairy parlour became the packaging and aging room. The bulk tank room became the cheese-making room, what Rhonda called the *caseificio*. A new milking parlour was in process when the whole thing came crashing to a halt.

According to Rhonda, the Alberta government wanted her to milk and make cheese year-round. Rhonda, steeped in the high-country lore of her mountaineer mentor, wanted to make cheese only when her animals were feeding on grass, for the best possible cheese. In the end, faced with quota costs and an intractable government bureaucracy, Rhonda pulled the plug on dairy cows.

She briefly considered water buffalo, but in the end, Rhonda purchased one hundred sheep. Rhonda's mentor arrived before the equipment did. "We had to throw milk out at first," she recounts. It takes ten litres of milk to create one kilogram of cheese. "We made twenty-five to thirty kilos of pecorino every other day for the first six weeks, until the milk supply dropped in July." They made feta, semi-soft pecorino wheels, blocks of pecorino layered with the farm's air-cured beef salami, and *fresco*, a pasteurized soft chèvre-like fresh cheese. Don Ronaldo was horrified that his protégée wanted to make pecorino infused with herbs, so Rhonda set that particular project aside until after her mentor returned to Europe.

At the end of grazing season, four months after her beginnings, the shelves in the aging room hold dozens of rounds of aging cheese, inventory on the shelf that Rhonda estimates as being worth forty thousand dollars. She proudly slices me samples of all her labours. "We are a good half million into this project. It's a big experiment. But we are not going to walk away any time soon." Not that they'll need to: in 2011, Rhonda and Brian were honoured as "agri-producers of the year" by the County of Vermilion River, and her cheeses are sold at the farmgate and at half a dozen shops in Calgary, Edmonton and Banff.

Pecorino

Most North Americans are familiar with pecorino Romano, a hard cheese made of sheep's milk. In Tuscany, pecorino is not aged as long, "so the end result is good for sandwiches, pasta and charcuterie," says Alberta cheesemaker Rhonda Zuk Headon. Ivory-coloured or a rich yellow, less salty than its Romano cousin, and at 40 per cent fat, Pecorino Toscano is perfect for a cheese platter, its olive-and-nut flavour complemented by figs, toasted nuts, cured meats or olives.

Pecorino Toscano is a designated DOP (*Denominazione di Origine Protetta*, or Protected Designation of Origin), the Italian version of the European Union's classification that protects and recognizes indigenous foods and food production methods. This guarantee of authenticity is applied to cheeses, wines and spirits. Thirty Italian and thirty-seven French cheeses receive a designation, in France called Appellation d'Origine Protégée (AOP), along with others across Europe. There is no such system in Canada.

Aging alters cheese, changing its texture and flavour as enzymes and microbes break down fats and proteins. Aged pecorino is ideal for grating onto pasta and risotto. Fresh, or *fresca*, it is good on bread for grilled cheese, or layered in pasta. The best way to enjoy any cheese is almost always au naturel.

Cheesemaker Jan Schalkwijk uses all his senses to produce award-winning Gouda at his Red Deer cheese facility and local-foods market.

Sylvan Star Cheese Farm Red Deer
Jan, Jannie and Jeroen Schalkwijk

"It is still a secret, what we do," says Jan Schalkwijk. "You cannot teach it. If you have not the feeling in your fingers, you never can make good cheese. Without the feeling, you learn the process, but you miss the point, how to tell what is the best, and what is not good." Jan knows the meaning of "best." His cheeses, made in a sparkling red and white building near Sylvan Lake, are award winners at national and international levels. "I can hear it, how the noise is—acid, growing inside."

Jan grew up in the Netherlands, where his father kept Holstein cattle and his mother made award-winning Gouda. He followed in her shoes, converting 1.5 million litres of milk into cheese annually. "Many rules and no room in Netherlands," Jan comments. He had no plans to make Gouda when he and his wife, Janeke (Jannie), followed their son Jeroen to Alberta, but observes, "I couldn't find any quality cheese I liked. So I started again in 1999, with milk from Jeroen's Holstein herd." In 2004, Jan's extra aged Gouda placed fourth at the world cheese championships; two years later, his smoked Gouda achieved the same accolade.

Sylvan Star's cheese is made with unpasteurized raw milk (legal in Canada if the cheese is aged for more than sixty days). The milk is "thermalized," heated to 60°C before the hands-on process begins. Some wheels are seasoned with herbs or spices.

In 2010, Jan and Jannie opened a new production building on the farm, with a large retail space, café and upper-level viewing gallery. I tour room after room, racks filled with rows and rows of wheels, cheese aging like fine wine—from mild, aged sixty days, to extra aged "Grizzly" (named for its bite and power), which is aged for eighteen-plus months.

The work is time-intensive and labourious. "On the day I make the cheese, I turn each wheel five times," Jan tells me as we tour the vast cheese rooms. "Then in brine for two days, and for eight days after that, it is waxed, half a wheel at a time, four times every side. Then, until it reaches sixty days, I turn it once a day. After that twice a week, until a year old, and once a week until it is sold." When he concludes, I have lost count of the times human hands touch each wheel.

At maximum capacity, Jan estimates he will have space to age one

Wheels of aging cheese are as much an investment as is quota, the Canadian supply-side system that controls cows' milk production and sales.

hundred thousand kilograms of cheese. Those wheels represent a significant investment in time and patience, and like the hands whose labours I try to inventory, go a long way toward explaining the high cost of aged cheese. Each wheel shrinks as it ages, losing up to 20 per cent mass and moisture. It takes time to make that leap to immortality.

Jan's cheeses continue to win gold medals, in 2006, 2009 and 2011 at the Canadian Dairy Farmers' Cheese Grand Prix. His sweetest victory came as a result of kismet. When a shipment of culture proved to be for Gruyère instead of Gouda, Jan went ahead and made Gruyère. To his utter surprise, it won the Canadian championship in 2009. "Making Grizzly cheese, that is normal for me, that is the standard, I know it for thirty years. New for me was the Gruyère."

After decades of cheese-making, Jan still eats his cheese every day. Medium Gouda for breakfast, old Gouda for a snack. "With wine, I have Grizzly, the best, the highest-quality cheese."

His ear and palate are still keenly attuned to his cheese. "For Gruyère, which uses a different culture than Gouda, I want to know how big the holes will be. You can't cut it open, you can't see in there. I hear sounds, my ear against the wax, the gas coming. You can hear that."

Milk quota

Regulations governing supply and demand saw the implementation of a Canadian milk production quota in the 1970s. Quota is a controversial, complicated issue. It evolved from supply-side market forces dictating that consumer demand constantly outstrip supply by dairy farmers' cows to keep milk prices stable for the cows' keepers. The quota itself has become a valued commodity, as scarce as hens' teeth, and costly. According to Julia Rogers, head of Cheese Culture, an Ontario-based cheese consulting and education business, acquiring dairy milk production quota costs $30,000 per cow being milked, and is strictly regulated, with each province allocated a set percentage. But there is no quota to be had: since its issuance, it mostly passes from family hand to family hand, and is rarely sold.

Milk processing is controlled by yet another quota, one that limits the amount of fluid milk pasteurized and converted into yogourt, sour cream, cheeses of specific types, ice cream, butter and powder. A third cheese quota governs importation from foreign countries, and is noticeable on our plates when importers trim their summer list to stockpile for the winter season. The permission to engage in business has become a valuable commodity in its own right.

To further complicate: quota applies only to cows' milk. There are not yet such regulations controlling sheep or goat milk, perhaps because, as Rogers comments, there are few dairy sheep and goat herds in Canada, and each animal produces a fraction of what cows yield. Smaller potatoes, perhaps, but for cheesemakers, dealing exclusively in goat and sheep milk removes one hurdle in their bid for cheeses to cross the Great Divide.

Canadians consumed twelve kilograms of cheese per capita in 2006, less than half of French consumption. It has never made any sense that a BC-made cheese that is safe to eat in Vancouver without a federal licence cannot legally be consumed across the border in Alberta. The major issues in federal inspection are the requirement for HAACP certification (a gate-to-plate tracking system that ensures food safety, wholesomeness and security) that includes temperature and time tracking during processing to eliminate contamination risks, packaging and labelling compliance, stringent building requirements, and the presence of a federal inspector in house, all the time or for a certain number of hours daily.

None of this begins to address the underlying issues of how our animals are kept, why raw milk is no longer safe to consume (after thousand of years of being a dietary staple!), or why raw milk cheeses are legally required to be aged a minimum of sixty days. For such a simple food, the leap from milk to cheese, from farmgate to plate, is fraught with complexities.

almost alsatian flambée

This seasonal bread makes a delicious appetizer or light lunch. Keep the toppings minimal, choosing the best local ingredients you can find. Alternatively, strip it down to basics: sautéed onions and garlic, sliced bocconcini, fresh basil leaves. Makes 4 10-inch (25-centimetre) rounds.

dough:
1 tsp (5 mL) yeast
1 tsp (5 mL) honey
½ cup (125 mL) hot water or milk
4 cups (1 L) all-purpose flour (or replace some of the all-purpose
    with whole wheat flour)
1 Tbsp (15 mL) kosher salt
½ tsp (2 mL) dried herbs (oregano, thyme, basil)
1–1½ cups (250–375 mL) hot water or milk
2 Tbsp (30 mL) cold-pressed olive, flaxseed or canola oil
2 Tbsp (30 mL) cornmeal (for the pans)

toppings:
2 red bell peppers
2 Tbsp (30 mL) olive oil, for pan
2 onions, sliced
6 garlic cloves, slivered
4 double-smoked bacon slices, diced
2 tsp (10 mL) minced fresh rosemary
3–4 fresh or sun-dried tomatoes, seeded and diced (see Cook's note)
1 cup (250 mL) slivered fresh basil
4–6 green onions, minced
3–4 sausage links
1–2 Tbsp (15–30 mL) cold-pressed olive, flaxseed or canola oil
kosher salt and freshly cracked black pepper to taste
2 cups (500 mL) grated aged Gouda
2 cups (500 mL) arugula

To make the dough, combine yeast, honey and water in a large mixing bowl and let stand until puffy. Add flour, salt, herbs and milk or water. Mix together with a dough hook or a wooden spoon until a smooth ball forms. Turn out the dough onto the counter and knead until smooth and soft. Oil the bowl, add the dough and turn it within the bowl so all surfaces are lightly covered with oil. Put in a warm place to rise. When the dough has doubled in bulk, punch it down without kneading. Divide in 4 and shape into flat rounds. Place them onto baking sheets lined with parchment and dusted with cornmeal.

Blacken peppers over an open flame. Transfer peppers to a bowl, cover and let stand for 5 minutes, then peel off the blackened cellulose, remove the seeds and slice. Heat the oil in a frying pan, then add onion, garlic, bacon and rosemary. Cook until soft and tender, and all fat is cooked out of the bacon. Add the tomatoes, basil and green onions. Poach sausage, covered, for 10 to 15 minutes, or until the juices run clear, then drain, cool and slice into thin rounds.

Preheat the oven to 400°F (200°C). If you have a baking stone, put it on the lowest rack and preheat for 45 to 60 minutes.

Distribute toppings (except arugula) on each round, drizzling each with oil, and finishing with salt, pepper and cheese. Bake for 12 to 15 minutes on the stone or on well-spaced racks until golden and crusty. Remove from the oven and evenly distribute arugula overtop the rounds. Cool briefly before slicing.

Cook's note: To seed tomatoes, cut each tomato in half along the equator line, then gently squeeze out seeds and juice.

Smoky Valley Goat Cheese Smoky Lake
Larry and Holly Gale

The acorn doesn't fall far from the tree, nor does it behave radically differently. Holly Gale, who makes goats' milk cheeses on the family farm in Smoky Lake, in north-eastern Alberta, learned to make cheddar and Welsh Caerphilly at her mother's side as a young woman. "Mom had a goat soap-making business in Beaverdale; she sold the soaps in the Okanagan. We made cheese and fudge, but soap was her business, and cheese was the hobby. Mom got into things too big," Holly says ruefully. "I was forever there helping."

Holly's husband, Larry, worked on a dairy for twelve years before he became a trucker, and he was supportive when Holly purchased six Nubian does and a buck in 2002 so she could make cheese and yogourt for their family. The herd grew to twenty-seven, and Holly could not sell or give away any excess raw milk. So she turned to books to learn how to make French-style cheeses.

She and Larry built a parlour and a milking platform on their nine-acre farm, and Holly tracked down cultures, buying an old-fashioned teak press along the way. She now makes bloomy rind Sainte Maure and Valençay, a washed rind Tomme, blue, chèvre, feta and farmers' cheese.

In her first year of business, Holly made enough to sell at Calgary's Kingsland Farmers' Market and several shops and restaurants. "But goat cheese is not a main commodity, it's a luxury item, and we were running in the red every month," she recalls. They decided to duck the rat race of large-scale production and Holly sold her goats for working capital.

She's currently milking eight does from her mother's herd and Larry has gone back on the road as a trucker. Holly, in her late forties, has set herself a pace that she can manage singlehandedly on top of caring for their son.

"Quota is set up to protect big industries, at the expense of—what? I won't even go there." But she remains firmly committed to producing for strictly local markets. During the winter season, her cheeses are available to members of the farm's cheese-only CSA and are retailed in Edmonton, Calgary and St. Paul in spring, summer and fall, as the does' milk production goes up.

Large-scale commercial success is more expensive than what the Gales are willing to pay. "No quota for goats' milk is a really big deal," Holly says vehemently. "If there had been quota, we couldn't have done it. We don't want to be big, and we were being pushed that way. My dream is to be able to help people so we can find small artisan cheesemakers in every community where it is inexpensive to start up. I am going to persevere, I love making cheese. It happens to everyone, we all face the same dilemma: Do you work off-farm?"

Old West Ranch Ltd. Mountain View
James and Debbie Meservy

For Brody Meservy, water buffalo are a fact of life. At Christmas, the little boy expects to hear that his favourite animals will take their place at the Christ Child's manger. At Halloween, when stories are recounted about a little white ghost who could only eat white things, Brody pipes up, clear as a bell, "Water buffalo cheese, that's white!"

"Yes, indeed," says James, his father, "not many families have that story." When the tall former dancer puts his *mozzarella di bufala* into my hands, he seems like a big kid himself, laughing and lighthearted, although he tells me in all seriousness that his cheese-making business, Canada's first farmstead water buffalo mozzarella cheese, is intended to conserve his wife's multi-generational family farm for their children. James doesn't come right out and say it, but the venture has the aura of a smiling gambler's almost desperate fling of the dice against the green baize of the craps table.

James, in his early forties, still moves with an athlete's grace, earned through decades of dance, including four years with Edmonton's legendary Shumka (Whirlwind) Ukrainian dancers' troupe. But he has traded in his red leather dancer's boots for farmer's gumboots in his persistent pursuit of a farm life for his children. He and his wife, Debbie, and their five kids were just scraping by on his income as a grad student in Texas when James, studying to be a geneticist, realized they were just too far from family. "If we are going to be poor while we have kids at home, there's a nicer place to be poor," he said.

In 2000, they came back to Debbie's family home, seven quarter

Buffalo mozzarella is so mild and tender that it is best served simply, with good tomatoes, olive oil and salad greens.

sections of land alongside Fish Creek, where the prairies meet the mountains in south-western Alberta, thirty kilometres east of Waterton National Park. James got a research job in a University of Lethbridge lab, and helped his father-in-law on the land, but decided that running cattle or making hay were non-viable. "I swore I'd never be a farmer when my parents lost their farm in Peace Country in 1986, but if we don't step up and take over we won't have a place to go home to."

James had made cheese and yogourt from the family Holstein for years, and briefly considered cheese-making, but milk quota was astronomically expensive. "We couldn't raise sheep or goats, nothing defenseless that is edible, we have too many large predators—grizzly, wolves—moving through our land routinely," James recounts. Their first alternative, musk ox, raised for their silken undercoats of wool, or qiviut, softer than cashmere, was a flop without the funds to build a harvesting facility. In 2007, with the bank breathing down his neck, James remembered water buffalo. He wanted to make real mozzarella.

James couldn't afford breeding stock from Darrel and Anthea Archer's herd of Murrah water buffalo at Fairburn Farm on Vancouver Island, but he did find a herd of eight cows in Vermont. No bull accompanied the harem—he was disqualified from entering Canada—so James was left holding the bull semen. "But I have a really hard time telling when the cows are in estrus," he says, laughing. "I do have a young bull. I just have to wait for him to grow up." More cows and a pair of bulls arrived in 2011, and James is building a federally inspected cheese-making facility.

James began to make cheese in 2010, using space in Crystal Springs Cheese, in Lethbridge. Halfway through a detailed explanation of how to pull the pasteurized and coagulated milk into cheese balls, he digresses into a long and funny tale of learning to outsmart water buffalo calves, offering them a substitute teat so he could use some of their mothers' milk for cheese. "Problematic," he says, "they roll like a crocodile if you touch their head when they are nursing." I am easily diverted, and want to know how James knows anything about crocodiles, but he carries on merrily, and finally concludes. "The chefs I sell my cheese to in Calgary were patient and so keen from the get-go. 'Oh, it's soft this week, okay, firmer next week, okay.'" It is easy to understand the chefs' acquiescence. Fresh, porcelain-pale mozzarella di bufala is a gift from the sea of Albertan prairie, one that should be shared with all, including small children and their white ghosts.

Fairwinds Farm Fort Macleod
Anita and Ben Oudshoorn

My conversations with Anita Oudshoorn invariably have the haze of motherhood overlaying them. It isn't just that she is the mother of seven, or that I often see her with a baby in her arms, or that, like many mothers, Anita tracks the passage of years by her children's births. It's that she and her husband, Ben, make their living through that most female of fluids—milk. "We got goats since Amy was a baby in 1995. She had issues with other kinds of milk, and we tried feeding her goats' milk. Back then, you couldn't find goats' milk very easily, so I got a goat doe, and we milked her." Soon Anita was milking half a dozen and had more milk than her growing family could use.

Goats have a five-month gestation period, and the herd rapidly increased to eighty. When Natricia Dairy opened in 1999, the Oudshoorns were one of several families milking and shipping their milk to the chèvre fromagerie in Ponoka. But it didn't last. Virginia Saputo, scion of the Montreal-based cheese family, and her partner, Mark D'Amato, made hand-ladled, surface-ripened cheese to the exacting specs of the French

Department of Agriculture, with an eye to export to the EU. Virginia made over a dozen varieties, including superlative Sainte Maure (cylinders traditionally rolled in ashes), crottin (a rough oval with an ivory rind), brie caprine, and Pyramide. The place quietly closed, to great disappointment and with no explanation, in October 2003. Anita and Ben were left afloat in a sea of milk.

The saving grace was that partway through the Natricia era, Anita had been approached by Community Natural Foods in Calgary to make yogourt. Anita put her foot down. "I told Ben I'm not doing this on my own, you need to get out of hogs and cattle, and we need land to raise our own feed; we need irrigation." The family relocated in 2003, to a larger piece of land west of Highway 2, on the shores of Willow Creek.

The farm and livestock were certified organic in 2008. Three hundred and fifty does produce up to six thousand litres of milk each week in summer. They still make yogourt, and Ben makes feta, luscious chèvre and aged cheeses, Gouda, Monterey Jack and Caerphilly. He's interested in making bloomy rind and washed rind cheeses as well, and Anita says they will give the results to their chef clients for appraisal. "Chefs are the best critics. It takes three or four years to develop a market," Anita explains. "Goat Gouda, people had to figure out how to use it. But feta! It's our biggest hit in the summer, because it's so good on salads, right?"

"Management is our job now. In 2003 we did it all; we were busy, but our kids are growing up too. We made the most money when we did it ourselves, but you have to keep going with growth and you can't keep up the pace on your own. Betsy, Gerda, Estella—our goats love us. They're all over us; they're sociable, playful animals. Smart too; they have their own characters. They are good therapy, and not boring." She adds, "Good does can be milked for ten or twelve years." Anita's a mother; she knows milk.

n is for navy beans,(great) northern beans, and other edible dry beans, lentils and pulses

Owen Cleland Bow Island

Dry edible beans and lentils, called pulses, just don't communicate glamour to most North Americans, unlike in the rest of the world. Scratch a menu in Toulouse, in southern France, and you'll find cassoulet, made with beans. Eat with South Asians and odds are good you'll be served lentils *and* beans. Few locally grown beans and lentils show up on Canadian store shelves, other than in shops that cater to an ethnic population. (As a new resident of Saskatchewan, I was overjoyed to find provincially grown peas and lentils on a shelf in a Persian shop in Saskatoon.) Even though bean growing in Alberta goes back to 1964, what Albertan beans remain in the province are sold in several stores in the "sun belt" where they are grown.

Nowadays, most Albertan beans are processed—dusted, polished and sorted by size—in plants belonging to Viterra, a huge corporation based in Regina, Saskatchewan. (The Alberta Wheat Pool became Agricore Cooperative when Alberta's and Manitoba's Pools merged in 1998. There was another merger in 2001, then, in 2007, the Saskatchewan Wheat Pool took over Agricore, privatized and became Viterra.) Then the pulses leave the country.

In 1964, a band of four farmers, including Walter Cleland, formed Alberta Bean Growers. Walter's son, Owen, grew beans for nearly twenty years, and is now Viterra's resident bean expert.

"Dad had first tried beans in the late '50s, and they didn't do very well," Owen recalls. "He started with seeds for longer-season varieties that froze out up here. Dad knew he had to get farmers involved or it wouldn't fly. He got three other growers in and formed Alberta Bean Growers. It was pretty successful, and grew to thirty-plus shareholders, and eventually got to the point where the business needed to expand and the farmers didn't have the cash. So they sold to the Alberta Pool in 1978. Back when he started, less than one thousand acres were seeded to beans in Alberta."

Great northern beans, second largest in volume grown in Albertan production, go to the Mediterranean rim countries, mainly Greece and Turkey, and some Arabic and North African ports. "Why don't Canadians eat more beans? We're not born and raised with it. We are still a meat-eating nation," Owen says ruefully. "Most pintos cross the border to the US's Latin American population in the southwest, and to the Caribbean and Africa. We keep maybe 2 per cent in Canada, and that's because of the influx of Mexican Mennonites as labourers."

Casey Koomen, a long-time bean grower based in Taber, explains, "The Mexican Mennonites were originally Mennonites from southern Manitoba's Red River Valley, where they created trades and tidy farms. But they didn't want to speak English or integrate, so one group seceded to Mexico, near Chihuahua, and another group went to South America. Some of the Mexicans came back to Canada a couple generations later, settled in the Taber area and work on farms. Our over-counter sales have skyrocketed: what we sell locally in southern Alberta, across the counter at our offices in Bow Island and Taber is up to twenty thousand pounds of beans. That's not much when you compare it to what we sell in total every year—one hundred million pounds of beans."

One hundred million pounds of beans. And it's almost impossible to find any of them in Alberta.

Leffers Farm Coaldale
Cornelius, Howard and Monique Leffers

Howard Leffers has seen a lot of smartly thought-out equipment emerge from the back roads where farmers live. "Most farmers are equipment people, not biologists," he observes. "To be a good farmer, you need to try to understand both: what kind of equipment, and what kind of bugs, do you need to make the soil happy?"

Howard and his brother Cornelius raise certified organic dry edible beans, alfalfa, carrots, beets and cereal crops in Alberta's irrigation country, near Coaldale, in the St. Mary River Irrigation District just east of

We all know what beans need

Dry-land farming lies just beyond the pale, in semi-arid near-desert where only the hardiest plants survive without access to running water or irrigation. In the 1970s, irrigation arrived in southern Alberta, and now, within the community of irrigated land, farmers "talk in circles." The pivot sprinkler system, a leading design in its field, situates a pivot in the centre of each quarter section to be irrigated. Each pivot, with a radius of four hundred metres, waters a circle, one hundred and thirty of a possible one hundred and sixty acres in each quarter. The four corners beyond the reach of the irrigation pivot are left dry.

Great northern, pinto, black and small red beans grow in a region that meets their specific needs. They need water from flowering to when the pod sets, and their roots don't draw from a metre deep down like chickpeas and lentils do. Beans need heat, as does corn, and are much more fragile to handle than you'd expect of such a robust food, but in fact, beans' thin seed coats risk breakage. These healthy and delicious legumes grow in a small area of land, in the irrigated "sun belt" of Alberta. Just east of Lethbridge, near Coaldale, running east along Highway 3 to Taber, Bow Island and Medicine Hat, no farther north than Vauxhall, and only five miles south of the highway, the growing season tops one hundred and ninety days, and the temperature is warmer here than in any other part of Alberta. Of forty-eight thousand acres seeded to dry edible beans in 2010, 80 per cent were great northern and pinto beans. Without companion planting, the beans claim all the heat and water. On the other hand, if it rains, the quality of a bean plant goes steeply downhill. It's a plant that needs babying, a plant that doesn't stake out territory. Weeds move in at the drop of a hanky, thinking the invitation is extended to stay and pull up a chair.

Lethbridge. "There's good information to be had from parents, and the older generation," he says. "You know, they say, 'Oh, we used to do this before we had that chemical.' Or, 'Be ready to do this if that happens.' We're losing our knowledge base here. The countryside is emptying." The son of Dutch immigrants, he and his wife, Monique, own four hundred acres, and began certifying chunks of it as organic in 2004, when they realized it was time to get big or get intensive.

Howard says he's "not a New Age Mother Earth type, but I am big on stewardship. The world is not going to last forever, but I think we should take care of what we've got. Our water is mountain runoff. The beauty of mountain runoff is that it passes through our hands, we don't pollute it, and it's just as clean when we're done if we do things properly."

Red and green lentils are staples in Mediterranean and South Asian cuisines, but are not widely consumed in Canada, despite the fact that 17 per cent of the world's lentils are grown in Alberta and Saskatchewan.

Crops are sown and harvested on a four-year rotation. Beans fall through the cracks in some years. They're chancy. They don't stand up for themselves when weeds come calling, and they don't command as high a price as Howard's good carrots and beets in three colours—red, golden, and striped candy cane, or Chioggia.

"Beans come if we can fit them in. I can afford to spend lots of money on carrots; they're a higher value crop than beans," Howard says. "Organic agriculture needs more labour, more management, and more 'local intelligence.' I mean that the farmer in a region knows his farm in a way that someone who takes over doesn't know, and in a way that just isn't really needed anymore if you use Roundup and seed treatment and pesticides and hope it grows." Conventional and monoculture farming isn't working on the Prairies, he says, any more than it works anywhere else. "There's accountant efficiency and there's real-life efficiency. The accountant's may be easiest to measure—humans aren't long-term thinking creatures—but the natural system doesn't do monoculture. So it's gonna get banged up some."

Howard sees flaws in comparing California and Albertan carrots. "It's not just about miles travelled and the costs. It's how they were produced. The cost of un-sustainability is not in the carrots, your grandkids will pay for that. And no one can put a price on that, not yet."

Howard always wanted to farm. He and Cornelius began with help from their folks. "This farm provides a touch-and-go living to our two families now," Howard says. "But if I have to get an off-farm job, then I think we should sell the place. If we still need to subsidize after we're up and running, there's something wrong with the system. Farmers are our own worst enemies. No one else keeps a job on the side."

Farm life is a great lifestyle, but it shouldn't be a charity. "We need to start figuring out ways to provide a living to farmers. You do that by providing products consumers want. Consumers nowadays want to know where it comes from, to talk to the guy who grew it." How do you put a price tag on that?

Saunders Farms Ltd. Taber
Jason Saunders

Canadians don't eat many lentils or chickpeas despite the fact that Canadian farmers grow 30 per cent of the world's peas and 17 per cent of the world's lentils, according to Pulse Canada. It's too bad. Both lentils and chickpeas are delicious, and have a long history of sustaining life. Lentils have been cultivated in the Near East since 7,000 BC; chickpeas were first tended in Turkey in the same era, and are grown in forty-five countries around the globe. Both crops have specific weather and soil needs that limit them to a very small region in Canada, mostly the arid land of Saskatchewan, extending into Alberta's southern "sun belt."

Pulses are also nitrogen fixers, meaning they partner with particular soil bacteria to take nitrogen from the air and turn it into a form that plants can utilize. Many farmers grow legumes like peas as "green crops" or "green manure" to turn back into the field for that reason. Others, like Jason Saunders of Taber, grow lentils and chickpeas because they turn over a good profit as well as a good field.

Lentils can be tough to grow. They have a shallow root system, don't take a tough stance against weeds, and are prone to disease, so they have to cycle through a three- or four-year crop rotation. Heat in the Taber region allows chickpeas and lentils to thrive, along with one hundred and twenty frost-free days, more than most of the province can claim. "Moisture could be an issue," Jason says. "Chickpeas need more stress to go to pod and seed, or else they keep growing and flowering. So if it stays nice, they won't seed." It's the seeds that are harvested, using straight combines that can also take down wheat and other grain crops.

Telling just when the pulses are dry enough to combine is done by the slightly arcane means available to a farmer. "We use telltale signs. The plant looks dead, and the seed is hard. Hard means dry. Lentils can be finicky. So you look, then test them with the combine."

Jason came back to the family farm in 1997 after completing two degrees in environmental soil and water, from a business approach and from the agricultural side. "I didn't plan to come back to the farm," he says wryly. "What changed? Life. I decided to give farming a whirl while my parents are still active. But I can't think of a better job. You have the freedom to run your own business. And yeah, you get out what you put in. But that's true of life." He laughs before carrying on. "You can have a pretty good family lifestyle, with time for the kids. It's enjoyable working from home, I have enough time to see them."

His business degree is applicable to everyday farm life, Jason says. "A farm is a business. Farming is a matter of how much you spend or don't spend. Yes, you can have awful years, but it's a matter of choices. You don't have to be farming twenty thousand acres to survive. I'm at eight thousand acres now, but I had a living already farming thirty-five hundred acres."

Jason's father grew peas and lentils in rotation with several other cereal crops before Jason took over. Receiving land from his parents made farming possible. "Land and machinery costs are too high for a young guy starting out. I wouldn't have done it if I had to start from scratch. My economics would kick in. I don't see that changing anytime soon, even Saskatchewan is rallying land values."

Land values around Taber are high beyond crop values, he says. Pressure isn't exerted by developing acreages or light industry in his neighbourhood, but it is elsewhere. "I am an optimist, but I'm also a realist. I don't see the world ending, so there's no reason to worry and live your life that way. But we do have to change our general habits. Everything you do has an impact."

One simple change that would carry a large impact, he suggests, is for Canadians to eat more chickpeas and lentils. That way, less of what Jason Saunders grows would be exported.

To soak or not to soak

No. Don't soak beans. There is no advantage to soaking beans, and lentils cook in a short enough time to preclude soaking them too. Soaking does strip out gassy carbon dioxide, but it also strips out many nutrients. It's easier, and tidier, to simply simmer pulses in abundant water, covered with a snug lid. Beforehand, rinse them and pick out any hard bits or pebbles. Remember that they quadruple in volume as they cook, so be generous with the cooking water and check them during the cooking process; there's simply no rescuing burnt beans or lentils. A pinch of fennel seed or aniseed added to the pot acts as a digestive. The older the legume, the longer it takes to cook, and the more water it needs to become tender, so don't be lulled by absolutist cooking times. Texture is the final and only arbiter to trust. Once the beans or lentils are tender, add sautéed onions and aromatics, pork, salt and a bit of vinegar to enhance flavour and spices. Freeze the extras for next time.

canadian cassoulet

I ate this succulent bean dish, called *porchas*, in a small traditional *comidas* in northern Spain, where it's made with shreds of *jamon Iberico*, the incomparable air-cured "black" pork haunch similar to Italian prosciutto. A similar dish, cassoulet, is eaten across the border in France, most notably in Toulouse, where cooks adorn it with duck confit, pork hock, smoked pork and sausages. Use beans or lentils and add what pork you like—smoked hock, ham, bacon or smoked sausage, prosciutto, capicolla or pancetta—if you didn't dare smuggle a jamon home on your last transatlantic flight. A drizzle of olive oil or cold-pressed organic canola oil adds richness and mouth feel. Expect this dish to take at least 4 hours of simmering time. Serve this hearty dish with simple greens, crusty bread and wine. Extras can find their way into quesadillas or the next pot of soup. Purée the tail end and serve as a dip or spread. Serves 10–12.

2 cups (500 mL) dried navy, cannellini or great northern beans
2 Tbsp (30 mL) olive oil or cold-pressed organic canola oil
4 garlic cloves, minced
2 onions, minced
1 tsp (5 mL) aniseed, cracked
2 bay leaves
2 cups (500 mL) shredded cooked smoked pork hock,
    smoked sausage, ham, pancetta or prosciutto
kosher salt and freshly ground black pepper to taste
1–2 tsp (5–10 mL) fruit-infused vinegar, or more to taste
extra virgin olive oil or cold-pressed organic canola oil to taste

Rinse beans and discard any stones. Cook beans in a large, heavy-bottomed pot at a simmer in generous amounts of water until tender, covered with a snug lid, adding water as needed.

Heat oil in a frying pan over medium-high heat. Add garlic and onion and sauté until tender. Add aniseed and bay leaves. Stir into beans when they are half-cooked, after about 3 hours. Add pork, cover snugly and continue cooking for another 2 hours or until tender. Season to taste with salt, pepper, vinegar and oil. Serve warm.

o is for oilseeds

Highwood Crossing Farm Ltd. Aldersyde
Tony and Penny Marshall

In a province known for its "boom and bust" oilfield ups and downs, an organic farm family in central Alberta has a rock-solid business from another kind of oilfield. "We created a market for a product no one knew about," Tony Marshall says reflectively, sitting in the family living room overlooking the Highwood River.

Tony's clothing sales background and his wife Penny's skill as a professional home economist led to the creation of a simple new product. Canada's first cold-pressed certified organic canola oil had an equally simple hook as "Canada's olive oil." "People were familiar with olive oil, and many had GMO canola in five-gallon jugs sitting in their pantry. The idea was to create a healthy niche," says Tony, who never imagined himself in Alberta's oil business.

Tony's great-grandfather, William Bevan Thorne, homesteaded beside the Highwood River in 1899. He and George Lane (one of the Calgary Stampede's "Big Four") introduced Percherons, France's celebrated draft horse, to Alberta. The farm he founded is now surrounded on three sides by encroaching suburbia, but the tree-lined river along the west side still runs clean and slow.

In 1993, during a sales trip to Germany, Tony observed a small oil press in action. On a subsequent trip, he bought one, came home and put up a building to house it on the farm.

"We looked at what people *weren't* doing: we pressed oil to order each week so we wouldn't have any sitting around; we used glass bottles; we used non-GMO organic seeds; no refining, no diatomaceous earth, no high temperatures. And we made the olive oil comparison stick, that canola and olive oils have similar profiles," Tony explains. "Canada's olive

In addition to cold-pressing certified organic flaxseed and canola to make highly-sought-after oils, Tony and Penny Marshall produce certified organic grain crops for their line of cereals and baking mixes.

oil" is floral and sunny, a monounsaturated fat, the highest in healthy omega-3 fatty acids.

"Slow" is an important word in the Marshall home. Tony, who is always nattily attired, is nicknamed "the godfather of local" in Calgary, although he is quick to joke about it—"It's just because of my age!"— before he turns serious. "I was fortunate to be at the Slow Food table as it was starting in Calgary. It gave a producer's face to the movement."

The golden version of "Alberta crude" is pressed on Tuesday, bottled on Wednesday and in stores or restaurants on Friday. Most is locally consumed, but PEI's Michael Smith of Food Network fame is a fan, and Quebec's Anne DesJardins has it on her menu at L'Eau à la Bouche in the Laurentians.

Beginning in 2001, Penny spent most of her time as a dedicated gardener, growing produce for Calgary's River Café. "I took it on because organic certification fees are so high, and we had to pay those bills. So while I browsed seed catalogues, and heard foodies talking about baby vegetables, I wondered, can we sell this stuff?"

They could. A string of River Café cooks worked alongside Penny, some barefoot and reflective, others at warp speed. "It was a wonderful time. Our hands were busy and present, we could visit or be quiet," Penny says. "Those cooks were so appreciative to be out of the city and their

Pandora's box: genetic engineering

It uses many aliases—biotech, genetically modified organisms (GMO), genetic enhancement or genetic engineering (GE), transgenic food, recombinant DNA technology. Whichever name it uses, this isn't traditional breeding as practised by centuries of farmers, selecting for favourable traits and breeding out the undesirable ones.

Genetic engineering changes plants and animals at the molecular level, artificially manipulating an organism's DNA by inserting genes or DNA segments from other organisms. Seeds can be engineered to be pest- and herbicide-resistant. Fruits and vegetables can be modified to slow their spoilage and give them a longer shelf life. It sounds like the stuff of dreams in a hungry world, but the downsides of GMOs outweigh the upsides.

Many environmental activists and organizations, including organics certification bodies, David Suzuki, Slow Food and Greenpeace International, are opposed to GMOs, and believe scientists still do not fully understand the impact GMOs have on the environment and on human health. *Earth Times*, an online environmental newspaper, posits that undesirable genetic mutations may lead to allergies and may impair nutritional value. Foreign DNA may survive in the human gut, and animals fed GM feed are prone to liver and kidney problems.

GMOs also fly in the face of biological diversity, long regarded as the key to global survival. Shrinking seed diversity gives greater control to corporations like Monsanto: the largest seed company in the world, it owns 86 per cent of the GM seeds sown globally.

Field studies show that herbicide use increases in GM crops: tougher weeds evolve and some plants become herbicide-resistant. Seed drift and pollen flow mean that GM seeds do not necessarily stay where they are sown, as witnessed by Saskatchewan farmer Percy Schmeiser's legal battle with Monsanto over Roundup Ready canola seeds that drifted onto his farm. (In 2008, an out-of-court settlement saw Monsanto pay for contamination costs.) Once GMOs are released into the environment, they cannot be recalled.

The Canadian Biotechnology Action Network (CBAN) lists four GM crops grown in Canada: corn, canola, soy and sugar beets. They show up in processed foods, tofu products, and canola oil, and in eggs, milk, meat and sugar. Look online for further details. Buy organic and unadulterated GM-free ingredients, avoid processed foods and read labels closely. In canola, for example, only organically grown plants are GMO-free, so only organically raised canola seeds make GMO-free oil.

stressful jobs." But the garden took a high toll and Penny reluctantly terminated it in 2008. "Human fatigue," she says, "and no long-term design planning on the natural side."

Producing a value-added product often costs more time, effort, labour and travel than is initially budgeted, Tony observes. "It's a huge amount of work, all long-term. Some agriculture people are so shy, and in this job it's painful to not be an outgoing person. You just have to be a different cat." The Marshalls don't follow the traditional farming model. They outsource additional seeds on a handshake from other organic growers, and Tony spends much of his time on the phone and computer, and making weekly deliveries.

The Marshalls point to a variety of factors that have made their niche successful: luck, good timing, proximity to a large city with a relatively large disposable income, strong service and marketing. "Of course it's slow," Penny says. "We can press between three and six litres of oil per hour." In Turner Valley, that's not even a blip on the radar. But on a farm beside the Highwood River, it's a gusher.

herbed honey vinaigrette

If this dressing separates, shake or stir it well before using on robust grilled vegetables, tender new greens and grilled fish. Makes about 1 cup (250 millilitres).

2 Tbsp (30 mL) mustard
1 tsp (5 mL) minced fresh thyme
1 tsp (5 mL) minced fresh tarragon
1 tsp (5 mL) minced fresh parsley
1 tsp (5 mL) minced fresh chives
1 Tbsp (15 mL) melted honey (see Cook's note)
¼ cup (60 mL) fruit-infused or mead vinegar
kosher salt and freshly cracked pepper to taste
¼ cup (60 mL) cold-pressed canola oil

Make the vinaigrette by whisking together the mustard, herbs, honey, vinegar, salt and pepper. Slowly add oil, whisking to emulsify. Taste. Add more vinegar if the dressing is too oily.

Cook's note: Melt honey in a small glass bowl in a microwave oven on low power or by placing the jar of honey in a small pot of simmering water.

p is for pork

First Nature Farms Goodfare
Jerry Kitt

Jerry Kitt's biggest surprise in 2008 as he wandered the miles of aisles at Salone del Gusto, Slow Food's Terra Madre companion food show, were the jars of Italian lard infused with herbs and caramelized vegetables. Jerry pastures Berkshire pigs in Alberta's Peace Country. He gathered up six jars after his eyebrows shot through the roof on the daily bus trip back to the monastery where he was billeted with three dozen other Albertans. "Pork lard? This good? Maybe this is what I should be doing as a value-added product," he said as he scraped the jar clean with a hunk of bread.

Jerry's certified organic farm, First Nature Farms, is an hour west of Grande Prairie, where the boreal forest meets the plains. Quaking aspens around him are first-growth, and the little-travelled gravel road just past his driveway takes the unsuspecting driver the final four miles to the BC border. Along with the pigs, he raises bison, grass-finished Angus and Galloway beef, pastured chickens and Merriam's wild turkey.

"When my wife, Sam, and I started out, our neighbours lent us tractors and gave us good advice. We learned one step at a time, cutting hay, then baling. It looked easy," Jerry says.

It wasn't easy. In the late 1980s, the family survived because both Sam and Jerry worked off-farm. Then they heard about holistic farm management, and learned to assign priorities to issues and values in their lives. Top-of-list was a peaceful existence for their livestock and themselves. The farm is rich in wildlife, so protecting the land and biodiversity had relevance too.

His animals live a humane, spacious life. The pigs indulge in naturally piggish rooting, wallowing and foraging. "It's odd to consider pigs as grazers, but they are," says Jerry as he leads the way through muddy fields to

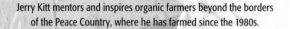

Jerry Kitt mentors and inspires organic farmers beyond the borders of the Peace Country, where he has farmed since the 1980s.

several small buildings in a separate field. The nurseries, he explains. Each building houses a recently farrowed sow with six or eight cute little piglets cavorting about her. "Don't pick one up or try to touch one," he cautions, "the sows will go on the attack to defend their babies." The impulse is almost irresistible: the piglets are spotted, solidly cute, curly-tailed and reminiscent of cuddly puppies. The sows, however, look lethal.

"We didn't sell much direct in the beginning," Jerry says as we head to the house for tea. "Even though we were raising organic animals, some went to the auction mart and we got commercial dollars for them, not what they were worth." Jerry, now divorced, makes a farm life without off-farm income. His meats are sold by several retailers and by direct sale, and at Edmonton's Old Strathcona farmers' market.

Late that evening, he methodically loads the trailer, calmly herding one pig at a time through the gate and along a chute. When the trailer has half a dozen animals in it, silence suddenly drops, as immense as the Peace Country night sky, and the animals peacefully take a nap.

Raising pigs in the field is perhaps simpler than raising kids on the farm. As with many farm families, the younger Kitts, now grown and gone, are not inclined to become farmers: Donovan is studying engineering, and Kari, according to her dad, was always a "girly girl" not much interested in farming.

"I see why the youth want to move off-farm. The average income of farmers doesn't compete with other professionals, and when our kids are growing up they don't see the benefits of a rural setting and raising a family on a farm, so they are pursuing different directions," he muses, tipping his broad-brimmed hat back from his tanned forehead.

There are other kinds of poverty. "I do this because wealth comes in other forms—knowing that I produce clean, organic food, and I enjoy spending time with animals."

Sunrise Farm Killam
Don and Marie Ruzicka

Don Ruzicka mentions his favourite writers as casually as we speak of our friends. "My goal is to hear [American poet-philosopher-farmer] Wendell Berry speak some day," he tells me. "We must become more intimate with our landscape. The word 'biophilia' means love of life, and of living things. Edward O. Wilson suggested in his book *Biophilia* that there is an instinctive bond between human beings and nature. I say that each of us is hard-wired to care for nature."

That explains why Don asks potential customers to visit Sunrise Farm. "Once they're at the farm," the sixtysomething farmer says, "when they run out of words and have had their fill of watching the pigs in the field, the creek in its green bed, the chicken in their pens, the cattle grazing, the solar water-pumping system, they can listen to the meadowlark and watch the bluebirds. The farm can tell the story." Canadians are within a generation of not being able to visit a farm and see how animals are raised. "Sunrise Farm is a vehicle to see that; we are cheerleaders for small agriculture and land stewardship. We all need a catalyst to do that."

The family's catalyst was an extreme one. Don and his wife, Marie, raised cattle and grain conventionally until 1995. They were deeply in debt. The problem was compounded by Don's struggle with Crohn's disease. "The operating loan had been going up by five grand each year,"

Seeing animals like this newly farrowed sow and her piglets on a farm allows Canadian consumers the opportunity to reconnect with nature, says Don Ruzicka.

Don recalls. "Crop failures, machinery that needed repair or replacement. It wasn't working. We knew we had to do something different."

"Something different" turned out to be an invitation to step off the agribusiness treadmill. "A holistic management brochure came in the mail one day. We went to Camrose and took the course. The rest is history."

Don and Marie divided their fields into smaller paddocks, fenced off the riparian areas to recover from cattle's incursions and grazed their animals rotationally, pasturing pigs and chickens, grass-finishing Angus-Galloway cattle. "A grass-finished animal without the right genetics and without a calm disposition will be tough, chewy and strong-flavoured. That's a huge factor in tender meat, how quiet the animal is."

Don is searching for heritage-breed Tamworth and Berkshire pigs, but he bought modern Yorkshire and Landrace pigs from a farmer seventeen miles away to support his faltering business. "They're long and lean, like a young Elvis, so beautiful and a pleasure to work with," he observes. After a summer of grazing, his pigs' livers are worm-free when tested after slaughter, a rarity among pastured pork that Don attributes to moving the pens twice a day.

Yogi Berra, the famous New York Yankees ballplayer and one of Don's preferred philosophers, famously said, "You can observe a lot by watching." It took nine years of watching for the birds to return to Sunrise Farm and to occupy the two hundred and fifty birdhouses Don built and posted. When climate change culminates in drought, he says, it will take longer to heal the damage done. "Grass-finished animals contribute more than omega-3s to our diet," he says. "When fossil fuel and feed barley go up [in price], what will grain-finished beef and pork cost?"

Don and Marie have evolved into environmental and farming leaders, admired for their clarity of vision and mystical wisdom, and sought after as mentors and public speakers. It's exactly what Wendell Berry meant when he wrote that when land is farmed in the right way, with respect and love, it has a positive effect on those who see it develop.

"The peace that people feel here is what I feel proudest of. That, and forty-five thousand trees planted since 2003, and this landscape reclaimed. We've gone one hundred and eighty degrees since we started farming," Don says. "It starts with one tree, one fencepost, one birdhouse. I see myself as empowering people, and yes, like Berry, I view myself as a simple farmer."

Cows and Fish

Don Ruzicka is a fan of Cows and Fish. The Alberta Riparian Habitat Management Society, better known as Cows and Fish, is a non-profit society that crosses the rural and urban divide. Water-influenced landscape, or riparian areas, are more than simply lovely. Riverfront land is biodiverse, productive, crucial to fish and wildlife, and integral to good water quality and supply. "Cows and Fish has done more for the health of riparian areas than anything else in Alberta," Don says.

Broek Pork Acres Coalhurst
Allan and Joanne Vanden Broek

Family ties are the strongest bond for Joanne Vanden Broek. She and her husband, Allan, grew up as schoolmates in the extended Dutch farming community near Lethbridge. "Our parents were friends," she says. "It's no big, glitzy romantic story: Allan's my best friend, after more than twenty years' marriage." For several years after they bought their farm in 1994, Allan worked part-time with his father-in-law, Bill Slingerland, and his brother-in-law Case. But Joanne remembers the time as dissatisfying. "When your husband is 'working out' from the farm, it seems you are split in two," she recalls.

In 2000, Allan and Joanne bought their first drove of piglets and raised them in conventional confinement. "We both grew up on pig farms; we knew we could start small and build the farm up," Joanne says. "But we sat in limbo until 2005." That's when their neighbour, organic dairyman Joe Mans, dropped by. "He said there are people in the US raising pastured pork naturally, and he told us about direct marketing."

Those words changed everything. The Vanden Broeks dug into research, and began pasturing their animals without growth hormones or antibiotics. For two years, Allan hired out the butchering, secondary meat-cutting, and sausage-making. The couple undertook face-to-face sales at the Lethbridge farmers' market, then began to deliver pork to restaurants and clients in Calgary.

Joanne realized that talking to customers was paramount. "Those one-to-one relationships taught us a lot—mostly about common sense, that we need to do what works for our customers. They tell us about flavours, and salt and spices. Our very first customer still buys pork from us."

By 2007, they saw the necessity for breed specialization and switched to Berkshires, a slow-growing but hardy heritage breed that can be traced back to Oliver Cromwell's era in England. Black-coated Berkshires were taken to Japan in the 1800s as gifts, and are revered as the *korobuta*, or black hog, with a reputation similar to Kobe beef's for tender, deeply textured and fat-rich meat. "Quality starts on the farm," Joanne says. "When we had more than one breed, it was easy to see. The Berkshires are the first ones out and nosing around under the snow for grass." It helps that they have a sweet, calm disposition that suits the relaxed outdoor environment. Their slower growth rate eliminates Berkshires from feedlot farming, Joanne adds. "Everything comes right down to the penny in conventional agriculture and the Berkshire doesn't fit that category."

To take charge of the butchery in an on-farm facility, Allan learned about meat-cutting from fellow pork farmer and retailer Greg Spragg, who had taken time off from his farm near Rosemary to attend a five-month butchering program at Calgary's SAIT Polytechnic. In retrospect, Joanne wonders if it might have been better, and maybe cheaper in the long run, for Allan to go to SAIT instead, but with nine kids to feed, it seemed a better bet to learn secondhand. "We eat our own mistakes. Learning on the job can be very costly—it's easy to make mistakes when you're teaching yourself," Joanne says, "and we're both still learning. There's so much to charcuterie; it's a specialized thing."

As of 2011, the farm produces, processes and sells an average of fifty Berkshires a month. "It's been a very large learning curve," Joanne reflects. "We were brought up to work hard, but we weren't businesspeople, we weren't butchers. We were farmers. That's what we value, the lifestyle of working together. And we live in next year country. We want to keep our kids on the farm. The bigger the family, the more you need each other."

charcuterie @ home: pork two ways

pork pâté

It's just as easy as making meat loaf to make this rustic pâté. Adding chunks of meat converts the finished dish into a rougher-textured pâté de campagne. Serve this rustic dish in slices on a platter with other charcuterie, grainy mustard and saskatoon berry or cherry compote, pickles and olives, crusty bread and good crackers. It freezes superbly, so consider splitting it in half and freezing a piece. Makes one 12-inch (30-centimetre) pâté, sufficient to feed dozens.

12 garlic cloves, minced
1 shallot, minced
oil for the pan
4 slices brown bread
½ cup (125 mL) whipping cream
2 lb (1 kg) pork shoulder
1 lb (450 g) pork fat
2 tsp (10 mL) quatre épices (recipe follows)
2–3 Tbsp (30–45 mL) ice water
¼ cup (60 mL) brandy, chilled
1½ cups (325 mL) mead, chilled
2 Tbsp (60 mL) maple syrup, chilled
½ cup (125 mL) chopped cornichons (small sour pickles)
½ cup (125 mL) chopped dried cherries
2 cups (500 mL) chopped mixed fresh herbs
    (parsley, thyme, spearmint, chervil, chives, tarragon)
1 tsp (5 mL) freshly ground black pepper
1 Tbsp (15 mL) kosher salt

Preheat the oven to 350°F (165°C). Sauté garlic and shallot in oil in a frying pan, then cool. Soak bread in cream in a small bowl. Coarsely grind meat and pork fat in a meat grinder, then grind the softened bread and add to the meat and fat. Mix in remaining ingredients. Sauté a small bit of the mixture and taste it for salt content and balance of flavours; adjust flavouring as needed.

Line a terrine or pâté mould (or a medium-sized heavy, enamelled cast iron pot) with plastic wrap. Put the mixture into the pan and vigorously whack the base of the pan on the counter a couple times to eliminate any air pockets. Cover with three layers of foil, securely sealed along the edge of the mould. Place the mould in a larger pan filled with very hot water so that water comes two-thirds of the way up the sides of the mould. The water prevents overheating and ensures even cooking. Cook the pâté to an internal temperature of 150°F (65°C), about 1½ hours, until the juices run clear when you insert a skewer in the centre. Put the pan on a tray, cover it with a wooden cutting board and stack several heavy tins on top to weight it down. (If the pâté doesn't protrude above the pan's edges, stack smaller cans directly on top of the foil.) Cool for a couple of hours, then refrigerate for at least 12 hours, still weighted down. Refrigerate uneaten pâté in the pan, wrapped, for up to one week.

quatre épices

It is called "four spices," but this classic seasoning for pâtés and terrines is usually made of at least five spices. The addition of dried herbs adheres to master chef Auguste Escoffier's classic interpretation. Makes about 4 tablespoons (60 millilitres).

1 tsp (5 mL) ground cinnamon
2 tsp (10 mL) ground allspice
⅛ tsp (0.5 mL) ground cloves
½ tsp (2.5 mL) ground cardamom
1 tsp (5 mL) freshly grated nutmeg
2 tsp (10 mL) ground coriander
1 tsp (5 mL) freshly cracked white peppercorns
1 tsp (5 mL) freshly cracked black peppercorns
2 tsp (10 mL) finely crumbled dried tarragon
2 tsp (10 mL) finely crumbled dried thyme

Mix together in a small bowl and store in a cool, dry place.

heather's smoked bacon

I learned how to make nitrite-free bacon from chef Heather Goulde-Hawke at District, in Calgary. Like Heather, I like the additional flavour that accrues when the rind remains on the pork belly. This recipe needs a smoker and a fruit wood, like cherry or apple. Makes 2 pounds (1 kilogram).

4 Tbsp (60 mL) maple syrup
2 lb (1 kg) pork belly, any ragged edges trimmed, rind on
4 oz (120 g) kosher salt by weight (about 1 cup/250 mL)
2 oz (60 g) Demerara sugar by weight (about ½ cup/125 mL)
1 garlic clove, minced
1 tsp (5 mL) minced thyme
cracked black pepper to taste

Smear syrup onto pork. Mix together remaining ingredients (the "cure") on a tray. Dip meat into the cure to thoroughly coat all surfaces. Shake off any residue that does not cling. Put meat into a waterproof plastic bag and store in the fridge for 5 to 7 days, turning it over daily. On the final day, rinse meat well under cold water and pat dry. Let stand, uncovered, on a plate or tray in the fridge to air-dry for 48 hours. This develops the sticky texture that allows smoke to penetrate. DON'T SKIP THIS STEP. Set up your smoker and smoke the pork belly for 6 to 8 hours over very low heat. Cool, cut into ½-pound (225-gram) pieces and wrap well for freezing. Freeze.

Cook's note: If you don't have a smoker, use your outdoor grill. Place soaked wood chips in a pan on one side of the grill. Set the pork on the other side on the rack over very low heat for up to 2 hours, turning several times.

Home charcuterie and the "salt box" method

Many cooks and diners shy away from making or eating processed meats because some contain (and require) sodium nitrite or sodium nitrate. These naturally occurring salts play an important role in meats that are ground and/or fermented or wet-cured. Nitrites are added to salt cures for smoked or dry-cured meats and sausages to kill bacteria, including botulism spores; to preserve the meat's pink colour; and to add flavour. But nitrites can produce carcinogenic nitrosamines, so meats containing them should be eaten sparingly. Nitrates, which slowly convert to nitrites, are added to long-curing meats and sausages that are cured but not cooked.

Many charcuterie techniques are within a home cook's range, but others are best left to technically inclined kitchen professionals. In either case, attention to details is mandatory. Without recourse to nitrites or nitrates, careful home cooks can make confit; fresh and smoked sausages; rillettes, pâtés and terrines; smoked and cured bacon and pancetta; duck prosciutto; and the Italian-style, air-cured beef called *bresaola*. The process always revolves around salting, curing and drying or smoking meat to deprive microbes, bacteria and mould of water, so cleanliness means safety. Use a bleach solution to sanitize meat grinders and stuffers, work surfaces, kitchen cloths and sponges.

Kosher salt is the heart of the matter. In dry rubs, the salt to sugar proportion is unchangeably set at 2:1, and that salt is there for a reason—the curing of the meat. Use an accurate scale to weigh salt and all other ingredients. Home cooks making charcuterie should follow recipes closely, without veering from specified salt (and nitrite, if used) content. Don't experiment. The easiest way to apply a salt rub is by using what is called the "salt box" method: smear the surface of the meat with all the wet ingredients, then mix together the sugar, salt and spices (and nitrites if required) and dip each piece of meat into the mixture to thoroughly coat all sides. Shake off any excess.

Hatching ducklings is one of the great pleasures of Gerwin Van Deuveren's work as a farmer.

q is for quackers (ducks)

Noble Duck Farms Nobleford
Gerwin and Esther Van Deuveren

"When I hit the Big 3-0 I thought I was getting old," jokes Gerwin Van Deuveren. It's easy to understand how a man who spends his time hatching and raising cute little ducklings would think of thirty as old—in our society, those fuzzy yellow babies are the intellectual property of kids and cartoons. But Gerwin is well below the average Albertan farmer's age of fifty-five, and the Pekin ducks he raises are wanted by more chefs and restaurants than he can supply regularly.

Gerwin was raised on a veal farm in southern Ontario's tobacco country of Tillsonburg. At seventeen, he visited Alberta. After raising calves for veal, working at the bottom of the chain in a feedlot at Picture Butte was no big shock. He came back to Alberta in 1999, and married three years later. He and his bride, Esther, acquired a farm yard—ten acres with buildings—near Nobleford, northwest of Lethbridge, and started feeding hogs and cattle. He says it was always clear to him that "farming was it. Once it's in you, it stays. I always liked working with animals."

Soon after, BSE shut down the cattle world. "When the Canadian dollar came up, it was all over for pork exports too," Gerwin recalls. "We decided we'd quit farming or look for other animals; I have no interest in grain farming."

Gerwin's brother-in-law Bert Vande Bruinhorst was raising lamb on his nearby farm, Ewe-Nique. In 2008, the two men got to talking about specialized animals, and Gerwin remembered that his grandfather had raised ducks in Holland. With a shrug to the universe, he thought, "Why not?"

"We spent as little as possible to get by; you just can't pour thousands of dollars into something that might not float," Gerwin says. They converted part of the barn into a "quackery" for their first one hundred ducklings, installing heat lamps and a lot of sawdust.

"Ducks are all about timing," he says reflectively. The hatchlings have a window of sixty hours after emerging from their shells when they continue to rely on the remnants of the yolk. After that, they need to be fed and watered like any other living thing. The little yellow balls of fluff arrived in a compartmentalized airplane box. "We spent a lot of time looking at them at first. Cute? Yes they were."

Finding a slaughterhouse wasn't easy. Ducks have a forty-nine-day feather cycle, and after they start to shed, their pin feathers are notoriously hard to remove. They have to be processed within days of that seventh week.

The day after the ducks are killed, Gerwin delivers them, plucked and chilled, to restaurants in Banff, Calgary and Canmore. At home, Esther cooks duck almost daily. "We love it, but I have only eaten my own," Gerwin says. "I don't know what other breeds taste like."

In January 2009, Gerwin got his first incubator. "The most rewarding part of the process is to hatch little ducks yourself," he says. Incubating duck eggs takes twenty-eight days. When the eggs are seven days old, Gerwin shines a flashlight's beam through them in a darkened room. He's looking for the fine lines that indicate the development of blood vessels inside the egg—any eggs that are clear are discarded. The incubator rotates the eggs by ninety degrees every ninety minutes, and subsequent exams by light show an amorphous shadow moving within the shell. On the twenty-fifth day, a tiny beak is visible in a sac in the shell. A silhouette inside presages arrival; a day or so later, the ducklings slowly start to peck their way out, through a crack that slowly spreads around the circumference of the egg.

The duck business is two years older than Gerwin and Esther's youngest of four children. Ducks are the family's primary source of income, and Gerwin hopes that his family will remain on the farm. "I don't like working with other people much. I hope to farm for the rest of my life. Yep, it's a tough way to make a living: a lot of work, and no money in it." But what it does offer—the chance to nurture ducklings into life amid the peace of a rural life—is enough to keep this farmer on the farm.

Greens, Eggs & Ham Leduc
Andreas and Mary Ellen Grueneberg

"I wanted to farm, and I dragged Mary Ellen and the girls along with me," Andreas Gruenberg says, his long face gloomy as he shepherds me around the greenhouse. Andreas and Mary Ellen met while skiing on Mont Tremblant. Their honeymoon was a cross-country drive to the Edmonton area in 1985. A graduate of McGill's agriculture and plant sciences programs, Andreas didn't have the readies to buy a farm, so he found related jobs, selling fertilizers, pesticides and forage seeds, and teaching. "When I hit forty, it was 'do or die,'" he says. They had just enough money put away for ten acres near Leduc in 1998, and Andreas continued to work off-farm.

With the farm, Andreas and Mary Ellen also assumed a pre-existing contract to raise pigs for Sunterra Meats, an Albertan company that raises, processes and retails meat. Soon after, Mary Ellen left her work as a

Duck duck yum yum

A *mulard* (or *moulard*) is a hybrid of Muscovy and mallard ducks, big-breasted and generous in the leg as well. It is a relatively rare breed, weighing about ten pounds (four kilograms) and is more commonly served in restaurants than in homes. Smaller and more readily available, Pekins, also known as Long Island ducks, weigh an average of four pounds (two kilograms).

A duck's meat is all dark, the colour generated by its cold-water home, working muscles and blood flow. It's rich, but leaner than many expect. The fat for which ducks are maligned is mostly in a layer between skin and meat, but some is intramuscular. A Pekin feeds only a few diners—its net edible weight after cooking is about half of what went into the pan. The rest is rendered out as fat. A mulard wears a fat jacket too, but it yields more edible meat, feeding six or eight, simply because it is so much bigger to begin with. It helps to remember that duck fat is 40 per cent monounsaturated, the good fat that is known to aid in the reduction of LDL cholesterol levels. Duck is best cooked in parts, after the fat is rendered out by a brisk sauté, skin side down. This limits the smoke in the house, eliminates the bulk of the grease and makes the most of two diverse types of meat. Legs are best slowly braised in stock or confited (simmered in duck fat); boneless breasts are superlative roasted to medium-rare.

Andreas Grueneberg hopes to eventually realize a profit from the greenhouse he built atop the barn on their ten-acre central Alberta farm. He and his wife, Mary Ellen, are involved in Slow Money, an offshoot of Slow Food in response to unimaginative treatment by chartered banks.

diagnostic imaging supervisor due to repetitive strain injury. "They were lean years," she admits. "The cash flow from the pigs helped."

Between 2000 and 2003, Mary Ellen tended flocks of laying ducks while home-schooling their daughters, Ariana and Megan. Gathering eggs became part of their daily physical education curriculum. They had pigs in the ground level of the barn and ducks upstairs, Mary Ellen recalls. "That's how Greens, Eggs & Ham started. We were raising vegetables to feed the ducks and produce the eggs."

A year after they began taking duck eggs and greens to farmers' markets, the couple participated in the inaugural Dine Alberta. This initiative partnered producers with chefs to raise the profile of local ingredients. But the lean years were not over. Sunterra moved its pork production to the US, and the Gruenebergs and thirteen other families lost their livelihood. Then the ducks stopped laying. "We had a lot of dead and sick birds from being sent the wrong feed, and we lost a lot of money." Mary Ellen began experimenting with meat breeds—Pekin, Muscovy and Grimaud hybrids.

"Everyone liked our greens, and kept asking, 'can you do it year-round?' So we transformed the upstairs barn into a greenhouse." Andreas re-roofed with polycarbonate and added lighting, tables and soilless potting mix.

Standing amid a dozen tables of Asian and mustard greens, Andreas says ruefully, "We did everything wrong initially." The well water's high sodium content built up in the tables, so now they truck in city water. When aphids developed, Andreas organized biweekly insect deliveries, alternating between thirty-five thousand ladybugs and five hundred wasps.

"For new farmers wanting to do something unusual, it is impossible to get money," Andreas comments. "Banks don't make decisions in house. They have their computer's drop-down menu and little boxes that fit conventional products. There is no room for vision, no personal story allowed, just the boxes. It's formula lending. Ducks don't even register as a farm commodity." He rolls his eyes. "The bankers could only compare us to broiler chickens—a large barn, quota, lots of birds. They understand that. There is no place for small-scale agriculture. Large-scale agriculture means profit potential is calculated in square miles instead of square feet."

Although the greenhouse has not yet made a profit, production and yield are increasing, with a goal of harvesting one hundred pounds of greens weekly, enough to supply twenty-four restaurants. On top are the duck and specialty duck products—prosciutto, sausage, smoked breasts and legs.

In late 2010, the greenhouse froze when a furnace switch failed. Reseeding and lost revenue were dispiriting, but not crushing. A private investor advanced money to cover the costs of duck flocks, and they implemented a "futures" plan, a modified CSA prepayment system that makes capital available. Ariana, now in her early twenties, returned to the farm to work. A weekly e-mail allows restaurants to pre-order, and the family attends markets in Edmonton and Calgary.

Mary Ellen sounds more hopeful than her husband. "Every time we think its time to give up, there's something positive. Customers tell us they appreciate us, or we get some media attention, or new customers. There is a need, and we are trying to fill a portion of it. We would like to fill more of it."

duck two ways

Duck is best treated as two animals on one frame: tender muscle-free breasts, and muscular legs with plenty of oxygen and blood flow. For the best results, abandon the hard-to-attain postcard image of perfectly roasted duck. Instead, sauté or roast the breasts, and braise or confit the legs.

mahogany glazed duck breast with fruit confiture

Sautéing first makes for lovely crispy skin and renders out much of the duck fat. A magret is one side of the breast, technically a half-breast. For a superb winter meal, garnish this with Canadian Cassoulet (p. 134). Serves 4.

4 boneless skin-on duck half-breasts (magrets)
kosher salt to taste
1 cup (250 mL) chopped dried fruit (black mission figs, apricots,
**   sour cherries, cranberries)**
½ cup (125 mL) port-style fruit wine, brandy, Armagnac, port or cassis
1 Tbsp (15 mL) butter
4 medium shallots or ½ onion, minced
2 Tbsp (30 mL) orange marmalade
water as needed
lemon juice to taste
1 tsp (5 mL) minced fresh thyme
hot chile flakes and kosher salt to taste
¼ tsp (1 mL) lemon zest

Pat duck breasts dry. Score the duck skin at ½-inch (1-centimetre) intervals in two directions without cutting into the meat. Salt lightly and let stand on a paper-towel-lined plate or tray for 1 to 4 hours.

Soak fruit in alcohol in a small bowl. Melt butter in a frying pan over medium-high heat and sauté shallots or onion until tender. Add marmalade and softened fruit and alcohol cautiously—it will ignite. Blow out the flame and simmer until softened. Thin with water if it becomes too thick. Add lemon juice to taste. Stir in the thyme, chile flakes, salt and lemon zest. Set aside.

Preheat the oven to 375°F (190°C). Sear duck breasts, skin side down, in an ungreased sauté pan for 7 to 10 minutes over high heat to eliminate extra fat and to crisp the skin. Transfer to a baking sheet, skin side up, and roast uncovered for 15 minutes for medium-rare. Remove from oven, let stand for 5 minutes, then carve across the grain into ½-inch (1-centimetre) slices. Use a long-bladed palette knife or spatula to transfer the slices, closely aligned, to a baking sheet. Re-set the oven to "broil." Brush the skin of the duck with 1–2 tablespoons (15–30 mL) of the fruit confiture and broil briefly. Serve with the remaining confiture on the side.

braised duck legs in white wine and star anise

Much as I love confit, or duck legs simmered in duck fat, it's just not a daily dish. Here's one that is. Made in advance and reheated, this makes any Big Meal or get-together a calmer one, and it makes a work-night suppertime splendid. Serve with a salad featuring a sharply acidic dressing, and mashed potatoes or noodles. Save one or two leftover legs for serving beside Canadian Cassoulet (p. 134). Serves 8.

ducks and rub:
8 duck thighs and drumsticks, attached, bone-in, skin on
1 Tbsp (15 mL) ground star anise
2 Tbsp (30 mL) grated ginger root
2 Tbsp (30 mL) olive oil or cold-pressed organic canola oil
4 Tbsp (60 mL) maple syrup
2 Tbsp (30 mL) finely grated tangerine zest
freshly cracked pepper to taste
braising liquid:
4 Tbsp (60 mL) flour
2 Tbsp (25 mL) reserved duck fat
1 leek, sliced
1 carrot, sliced
6 garlic cloves, sliced
1 cup (250 mL) aromatic dry white wine
2 cups (500 mL) brown chicken or veal stock
1 whole star anise
orange zest, 1 long piece
hot chile flakes to taste
kosher salt and freshly cracked black pepper to taste

Smear duck legs with rub ingredients. Transfer to a large bowl or tray, cover and refrigerate overnight, if time permits. If not, let stand for at least 1 hour.

Preheat the oven to 300°F (150°C). Sauté duck legs in a dry sauté pan (no fat added), skin side down, until well-browned, about 10 minutes. Drain and reserve the duck fat. Transfer legs, skin side up, to a heavy-bottomed braising pan. Sprinkle with flour. Set aside.

Heat reserved duck fat in the sauté pan and add leek, carrot and garlic. Cook until tender and beginning to brown. Add wine to deglaze, scraping up any browned bits, then add stock and bring it all to a boil. Pour liquid and vegetables over the duck legs in the braising pan. Cover duck and sauce snugly with parchment paper, pressing it right onto surface of meat and liquid, then fit with a snug lid. Braise for 2 to 3 hours, or until meat is tender. Season with salt and pepper to taste. Transfer meat to a large bowl and cover loosely with foil. Skim and blot any excess fat from the liquid in the pan, straining out the solids if you wish, then return meat to the pan. Reheat and serve hot.

Rillettes

Save a few cooked legs to shred into rillettes. Though traditional rillettes are made from pork, duck or goose legs make a nice variation.

To make rillettes: Begin with duck legs and a rub, elaborating on the flavourings for the rub at will: sweet spices such as allspice, star anise or aniseed, sparingly applied, make a good counterpoint to the richness of duck, as do ginger and Szechaun pepper. Cook meat as for braised duck legs above. When cooked and cooled, pull off and discard the skin from the duck. Pull the meat off the bones, saving the bones for stock. Use two kitchen forks to shred the meat and pack it into several ramekins. Top each one with enough melted duck fat to completely cover the meat, and chill. Age the rillettes for at least 3 days before serving. Serve rillettes in an uncomplicated fashion, alone or with pickles, mustard and good crusty bread, or as part of a charcuterie platter. Rillettes keep, covered, in the fridge for up to a month, and freeze well. One duck leg makes one ramekin of rillettes.

r is for roots

New Oxley Ranch Claresholm
Jackie Chalmers

"Garlic is something you plant and weed on your hands and knees, up and down one hundred times a day. My uncle says it's work all done on your prayer bones," Jackie Chalmers says. Her grandparents farmed at Millarville in 1905 after they arrived from Ontario, and as an adult, Jackie wanted to play in the dirt. But a change in Millarville's demographics drove her away from the family land. "It's too crowded there for me, there's been a big change. I shudder if my gran was to come back," Jackie says of Millarville's former farmland, now fenced and padlocked in front of mansions. So she moved to a ranch near Claresholm, south of Nanton.

After watching a mule deer calmly chewing on a flower from her planter just outside the kitchen door, Jackie knew that whatever she planted, she'd be sharing it with the deer. Then she remembered, not for the first time, the acumen of her family. "When they grew their garden west of High River, Aunt Ruth planted garlic between the rows of vegetables to deter deer," Jackie says reflectively. Jackie planted her first trial crop in the fall of 2008: fifty bulbs, each separated into cloves, of a particularly delicious Italian garlic variety. Would the deer like it too?

In 2009, after harvesting the trial crop, Jackie walked to the end of her long rows, now seeded with seven thousand bulbs of garlic. She saw that the muley had nibbled three inches off the top of one plant. It never came back. Jackie harvested a thousand pounds of hardneck Music garlic in 2010. With a viable cash crop, all spoken for by Calgary Co-op, restaurants and private individuals, Jackie began to examine ways to add value to her primary crop. "I can rent a kitchen; production is the easy part. Then there's sampling, marketing, labelling, packaging, distribution—I have to assess my abilities and my age."

Jackie Chalmers planted a commercial garlic crop after a neighbouring
muledeer declined to eat any of her test crop, harvested in 2009.

Jackie relied on her Aunt Ruth's old world wisdom in her rural life
while she raised her children, Jennie and Paden. When Jackie embarked
on a diversification plan in 2010, a pilot CSA with twenty-eight subscribers, she relied on Paden to do much of the heavy work. "It was Paden's
Pilot Pastured Poultry Project, really, more learning than money." Her
teenaged son earned his fee, hauling water, moving the birds daily to fresh
grazing. In 2011, the chores become Jackie's exclusively when Paden went
to Olds College. "At my age, in my late fifties, the desire is there," she
admits, "but the body is the issue."

The other issue proved to be choosing the right breed of bird. A trip to
Slow Food's Terra Madre and Salone del Gusto in September 2010 evolved

into an Italian research tour. On a Piedmont farm, Jackie and colleague Michelle Malmberg located the Gallo Nero, a famous breed they had seen photographs of in Slow Food's Presidia of endangered species. "It was the Meryl Streep of birds," Jackie says excitedly, "big, robust, beautiful, not all breast, not the Pam Andersons I had tried to raise at home."

"The more we can embrace this idea of raising food the old-fashioned way, and being financially rewarded for it, the better off we will all be. Living on the land in my generation, we had a full, rich life but nothing extra. I hope one of my children will take up life on the land, but they have to want to, it's gotta be hard-wired, and I want them to have a life quality similar to contemporaries who choose a mainstream job. If we ever get our kids back on the land, Slow Food will be the impetus. Watching the young people at Terra Madre was so inspiring. I want my kids to feel that excitement of how to make a difference."

Eagle Creek Farms Inc., Bowden SunMaze and Eagle Creek Seed Potatoes Bowden
John and Rayell Mills

When John Mills tours children through the sunflower maze on his farm, his face ignites, as brightly lit as his sun-bleached hair. In his early thirties, John and his wife, Rayell, are raising the fifth Mills generation in the house originally occupied by John's great-grandfather after the First World War; during renovations, John found a 1930s postcard addressed to his great-grandmother behind a fir baseboard.

When his dad said, "You'd be an idiot to farm like I have!" John went away for eight years. After a year visiting his mother's family in England, he worked in Europe, then Australia, and capped it with two years at Nova Scotia Agricultural College in Truro. In Truro, a u-pick flower farm and maze caught his attention. He transplanted the idea to the central Albertan valley where he was raised, growing Asiatic lilies and bright all-yellow sunflowers for Innisfail Growers. But flower sales plummeted with the downturn in the economy of the late 2000s.

The virtue of Slow was driven home to John in 2008 when he attended Slow Food's Terra Madre as part of the international "youth" contingent. Although he is part of a traditional mixed farm, the younger Mills does

In addition to a large CSA, John Mills welcomes visitors to his multi-faceted farm's sunflower maze, located on land farmed by his great-grandfather.

not view himself as part of mainstream agriculture. "I'm no activist. I farm with likeminded people to develop systems and models, to share alternatives, not to say, 'you've been doing it wrong.'" Going to Italy "solidified my goal of being a sustainable farmer," he says. "At Terra Madre, I saw the whole world, but realistically, changing an established situation is not as easy as it sounds. I have to take small steps." He remembers three Italian dishes he ate with the clarity of a Renaissance painting. "Hundred-year old balsamic vinegar like candy! Pickled garlic. Freshly made pasta with pesto. I can still taste them. The intensity of those flavours is what I want to copy on my own farm."

Since his return, John has simplified his approach. "I see it isn't just me who wants to revolutionize the world. I don't have to do fancy marketing; I don't have to change minds. I just have to grow good food." In 2010, he inaugurated a CSA on Eagle Creek, with sixty-six member families. The number increased in 2011, to a whopping two hundred families, and he's added a flock of meat chickens to extend the CSA season into winter.

John utilized heat-trapping practices to grow his first watermelon, honeydew and cantaloupe. "The first year, I seeded directly outside," he explains. "Now I put them in transplants, so they're started indoors." He is investigating the Lethbridge Research Station's methods of planting fragile crops inside four rows of corn planted in a square buffer. "It keeps hot air in the field," he says. "It is not the day's heat that is at issue. It is our cool evenings." Those cool nights make for great root vegetables with crunch and sweetness.

John has encouraged his father to move away from mainstream grain and cattle farming. They've adopted on-farm tourism, building mazes in corn fields and sunflower stands. Mazes are ancient spiritual forms of renewal, and for centuries have been used for prayer, ritual, initiation, and spiritual awakening: blind alleys and dead ends offer many opportunities to practice humility and perseverance, to know when to stop struggling and accept help, and to master the inner difference between a dead end and a bend in the path. Often, a walk through a maze mirrors the path through one's life. When the three-acre sunflower maze is in bloom, it forms a six-foot-tall sea of yellow towering quaking light whose susurrations are the ideal audio background for a slow stroll.

Twenty acres of seed potatoes—including Irish Cobbler, Kennebec and German Butterball—and John's mail order business remain the mainstay. But the potatoes themselves ignite that blazing grin. "It's my most passionate crop. They taste good, anyone can grow them. Kids make Mr. Potato Head stamps at our lily festival. They grow in different colours." He picks up a handful of red, yellow, white and purple spheres and tosses them into the air. "You can juggle them."

Poplar Bluff Farm Strathmore
Rosemary Wotske

Rosemary Wotske had her epiphany early. Her parents had left the family farm in the 1930s when times were tough. "We visited farm friends when I was twelve, and they had to physically put me in the car to go home," she recounts, her brown hair, the same colour as Irish Cobbler spuds, blowing in the breeze.

In 1985, in Pakistan with her husband Dean Giberson, it was time to act. "I turned that magic thirty-five, and I just knew, if I'm gonna do this, I gotta do it now," she recounts. "We'd agreed he'd work in External Affairs for fifteen years, then we'd do what I wanted to do." Rosemary returned to Canada alone—with South Asian recipes that enhanced her already considerable rep as a good cook—and enlisted her parents' help to buy a quarter section of land near Strathmore, east of Calgary.

Buying a farm, any farm, earlier had seemed unattainable because of the enormous cost of land. Instead, Rosemary had achieved a Bachelor of Science in physiology and biochemistry, and a Master's in genetics after her wedding. The couple went to Moscow and Pakistan. Then Rosemary commuted, from overseas, then from Ottawa, so she could spend summers on the farm.

Her first crop was an acre of asparagus. In 1989, she took the spears to restaurants in Strathmore. Only one chef recognized it. None bought it. Rosemary went next to the Millarville farmers' market, where she observed consumers frequenting tables laden with arrays of colourful produce. From asparagus to zucchini, she filled her beds, making preserves from anything that returned unsold from the market, and added animals—chickens, eggs, turkeys, hogs.

The farm was certified organic in 1998, just as Rosemary admitted she was getting tired. "Dean kept good records, so we examined them. Potatoes always did really well for us, so we specialized, thirty spud varieties, including Ratte, Agria, Rose Finn apple, banana, Kennebecs, Roko and Desirée, plus heirloom carrots and beets."

Scientist-turned-farmer Rosemary Wotske, right, has become a friend and generous source of growing information to the many who enjoy her organically produced potatoes and root vegetables.
PHOTO: KAREN ANDERSON

When several Calgary chefs approached Rosemary, asking about buying her produce, she was skeptical. "I had closed that door," Rosemary says acerbically. "But these guys were positive. They paid their bills. What more could you want?" The vegetables thrived; the marriage ended; a business partnership didn't work out. Then in 2008, Cam Beard, a former trucker, approached her, hoping to realize his dream of farming. In 2010, he harvested his first Poplar Bluff carrot crop.

"I wish I could say I was so smart, but it's easy in the best place on the planet to grow roots," Rosemary explains as she serves carrot *halwah* flavoured with green cardamom. We are sitting on the swing in her backyard. She points west, her skin russet in the late light. "Heavy soil. Minerals that provide nutrition. And this crazy Palliser climate." Alberta's warm sunny days, followed by high-altitude cold nights, produce robust plants. "The plants produce sugars in the presence of light, then convert the sugars to growth at night. It's a temperature-dependent enzymatic process. But the plants never use up as much sugar as they have made during the day so that's what we taste."

Rosemary's business revolves primarily around restaurants, with thirty weekly restaurant kitchen deliveries, and an equal number of now-and-then clients. Her biochemistry training is finally relevant. "I'm interested in how a spud starch changes, and in mouth feel, and do you want slices intact when you're done or all blended. I like that fun and experimentation, matching a chef with a spud he'll be happy with. I get to share their creativity. I always come out pumped. And full." She grins, patting a round midriff that could be a Kennebec.

When Rosemary narrowed her spud selection to twelve varieties, it wasn't hard to decide. "Agria is the best, after them there is no other yellow." And despite the fact that russets were prolific, their yield so heavy they would stall the chains on the harvester, they were ditched. "Russets taste like dirt," Rosemary says. Dirt is hard to sell as a flavour profile. What isn't hard to sell is Rosemary's rural life. "I think it's the open spaces and things growing everywhere. And the quiet, the animals. I plan on dropping in my traces. I'm incredibly lucky. I do what I love. I can't imagine why I wouldn't want to die here."

Lund's Organic Farm Innisfail
Gert and Betty Lund

It was the French painter Paul Cézanne who wrote, "The day is coming, when a single carrot, freshly observed, will set off a revolution." In Innisfail, that day arrived in 2007 without fanfare.

Gert and Betty Lund's organic carrots are a mainstay in many Albertan homes. My dog has eaten his weight in those crunchy sweet carrots, as have my sons. The Lunds, who farm just north of the Innisfail town line, are pioneer organic growers in the province. They arrived from Denmark in 1981, and their first farm in the Knee Hill Valley, twenty kilometres east of Innisfail, was certified in 1984. They moved to their current sixty-acre farm in 1992 for easier access to the QE2, the highway that is the main artery of Alberta. Ironically, that proximity may cost them the farm.

In 2007, the town of Innisfail, eyeing future development, annexed ten quarter sections of land north and west of the town's boundaries. Part of the annexation package was the Lund farm.

Gert Lund's organic carrots have become emblematic of the farmland annexation issues that loom larger as Albertan urban centres spill over into farmland.

"Stand in the way of progress? You'll get run over. But if I was twenty years younger, I'd get a shark lawyer with big teeth. At this stage in my life, is it worth it?" Gert says, sighing and running a hand over his balding head as he watches his dogs help themselves to carrots from the washing shed's buckets. In 2011, Calgary chef and local foods activist Wade Sirois attempted to assemble a consortium of owners to purchase Lund's Organic Farm to keep it going as long as possible, but Gert and Betty had not decided on a course of action. They don't know of any certified organic land available in the vicinity. Gert is pushing sixty, and although they have bought land in the Similkameen Valley, in southern BC, he says he is too old to start again.

It takes three years to transition conventional land to organic production, plus another five years to realistically learn how a piece of land responds, to nudge it into optimal production. The corridor along the QE2 highway will eventually be a mass of residents, with accompanying high-speed rail and electrical transmission lines. The drive north to Edmonton may eventually happen without the chance to view a single piece of open arable land under cultivation.

Dale Mather, Innisfail's former chief administrative officer, tells me by telephone that farmers like the Lunds are free to sell or continue farming their land until such time as development begins. It could be five years, or forty. "Innisfail allows annexed farmers to remain on a rural tax scheme until their land is developed or a subdivision is put in place," he adds, and wonders out loud why Calgarians would be interested in a small town's municipal development plan.

"Something to do with carrots," I tell him.

The town of Innisfail has 7,883 residents in 2011, including town councillor Jason Heistad, his wife and their three young daughters. "We don't want to be like the lower mainland of British Columbia," Jason says earnestly. "I am sure they didn't think that all that farmland in the Fraser Valley would be eaten up by paving. It is some of the best farmland in the country." Heistad sounds as ambivalent as many Albertans. "The town is aware of the concerns of farmers and is addressing them," he says, "but growth is a no-brainer." It's true. Growth is a no-brainer, except for the one small wrinkle of farms converted from carrots to concrete.

Despite a teenaged motorcycle accident, degenerative disc disease and colon cancer, Eric Deschipper loves the life of market gardening vegetables.

Red Willow Gardens Beaverlodge
Eric and Carmen Deschipper

Eric Deschipper was fifteen when he and his motorcycle were hit by a truck in 1981. Four surgeries and two years later, he received an insurance settlement and went looking for land. Red Willow Gardens was a market garden with half an acre of raspberry canes and an acre of vegetables in the lea of Red Willow Creek, carved beneath the escarpment and the creek itself. Eric thought it was a beauty. He and his parents co-purchased the land in 1984. But "God willin' and the creek don't rise" was not written into the deed.

"It had the advantage of water," Eric, now in his forties, recounts. "But it was way out of town, in a farming area, where everyone had their own garden." The closest farmers' market was in Grande Prairie, an hour's drive east. The deal was struck just in time for seeding. "We put in seventeen acres, but it was all new and we started late, so we had to buy nursery seedlings that first year, cold-climate vegetables—cabbage, cauliflower, broccoli, kale. And we planted roots," Eric says. "That year, we learned what doesn't sell."

The following year, they attracted the attention of a northern crop specialist who provided them with seed for the Nantes 616 carrot, at the time still an experimental variety. They planted three acres. "They were fragile, those Nantes, we had to hand-harvest them, so they meant more work. But once people got a taste of them, they were a huge hit," Eric says. Eric and his father logged tamarack trees and milled them into hardwood boards for a root cellar. To make ends meet, his father found work as a school custodian and janitor, and Eric worked on the oil rigs, where he met Carmen, who was cooking for the crew. Carmen gave birth to their first child in 1988, and a year later, they built their dream home on the Redwillow River.

In 1990 the river flooded after three days of non-stop June rain before the snowpack came down from the mountains. "The river rose so quickly that it was not a fact of how high but its speed," Eric remembers. "It tore a new river channel, tore land away from the farm, and flooded not our home and yard, but the garden itself. Three acres of land suffocated." The year's crop gone, Eric went back to the rigs.

Land use and farms

Between 1996 and 2006, the number of Albertan farms decreased by 9,576. Calgary is surrounded by lost farmland, and the city's arms have embraced much of the rich land around the city. Drive west, and observe Springbank's remaining farms stitched side-by-side to suburban sprawls. Bearspaw's high-bluff view of the big bend in the river, formerly a rural sight, is hemmed in by McMansions. Cochrane and Okotoks are minutes from Calgary city limits. The racetrack and mall development in Balzac have buckled up the narrowing agricultural belt between Calgary and Airdrie. The annexation of Airdrie may soon occur; the Calgary Airport Authority already operates the Airdrie Airport. Farther north, the farmland annexed by Innisfail is slated for residential development.

Data released in the May 2006 Canadian census shows that nearly 2,500,000 Albertans—73 per cent of the province's population—live in the "Calgary–Edmonton Corridor," the fertile 400-kilometre band of land that connects the two cities, including the cities of Airdrie, Wetaskiwin, Red Deer and Leduc, and the growing towns along the way. Eighty-one per cent of Albertans are urban dwellers; a mere 71,660 of the province's three-million-plus population are farmers.

In Edmonton, along the north side near the river, market gardeners like the Kuhlman family have spent decades staving off development in some of the most fertile land of the province. Kuhlman's Market Gardens & Greenhouse began in 1962 when Dietrich and Elizabeth operated a u-pick market garden. The present-day business, which feeds thousands of Edmontonians, consists of a garden centre and three hundred thousand square feet of greenhouses. The current generation, Anita Kuhlman McDonald, her husband, Doug, and her sister, Angela Kruk, and hubby, Dale, are resigned to the inevitable.

"The city encroaches," says Anita. "The Anthony Henday freeway is going through my parents' old house. That land was sold to the government in 1980 with ring road plans, but they never tell you when. We rented it for years, and are losing land base because of a utility corridor. We are looking to maybe downsize the garden or shift to only growing for farmgate sales and farmers' market." Anita thinks that the foursome must change or cut back. "We will be here for a while yet, but I don't think anyone is coming to say, 'we want to buy your land' so we can retire."

In 1994, he and Carmen bought out Eric's parents. They discontinued the greenhouses and Eric quit working the rigs. "All that other work busied us, so we missed the point that our money was in the carrot seed, and why leave them in the field?" Although he continued to grow other vegetables, the Nantes 616 became the farm's flag-bearer, until the seed variety disappeared. Eric tried to seed-save, overwintering extra acres of carrots in the ground, but the northern latitude's deep winter cold meant most of the carrots froze. Eric has planted other varieties, but he still misses the 616.

They opened a teahouse, where Carmen's carrot cake became so popular that it forced another exercise in focus: add a café and hire cooks, or do what they do best? They chose the carrots.

Eric was diagnosed with degenerative disc disease in 2006. "The doc told me to sell. I was heartbroken; it was our dream, I raised my kids here. We listed, but by April 2007, very few people looked. They didn't want the business or its workload of forty-two acres of land with two houses, a new store and a new root cellar. So I realized the good Lord wants us to keep feeding his people." Eric brought in Mexican staff and converted the teahouse into a staff residence.

In late 2010, five months after planting three thousand strawberry plants, and just back from a family vacation, Eric was diagnosed with colon cancer. Even as he sat with his feet up and awaited surgery in early 2011, Eric thought about eight acres of carrots waving their green flags in the air. His staff put in the crop. "Every year the challenges are met, one day at a time. Next thing you know, the year is done, and the root cellar is full. When you are a farmer you learn patience."

Grilling or roasting enhances the sweet nature of many vegetables, from asparagus to zucchini. Root vegetables, and carrots in particular, are best cooked in such a simple style.

grilled or roasted roots and veggies with minted yogourt

Roasting root vegetables in winter is THE best way to eat them, in my opinion. Use the grill when the weather is fine. Follow the seasons: carrots, squash, cauliflower, beets, parsnips and turnips in winter; asparagus in spring; peppers, tomatoes and eggplant in summer and fall. Serves 6.

1–2 Tbsp (15–30 mL) olive oil or cold-pressed organic canola oil
6 carrots or parsnips, split lengthwise
1 bell pepper, sliced into strips
1 bunch beets (see Cook's note)
⅛ head cauliflower, cut into small florets
1 cup (250 mL) batons of squash or turnips
kosher salt and freshly cracked pepper to taste

dip:
1 cup (250 mL) gelatine-free yogourt
2 Tbsp (30 mL) minced spearmint
1 Tbsp (15 mL) minced chives
2 Tbsp (30 mL) cold-pressed canola oil or olive oil

Preheat the oven to 400°F (200°C). Place all vegetables except beets in a thin layer on a parchment-lined baking sheet, and dress in oil, salt and pepper. Roast in a hot oven, 20 to 35 minutes, stirring several times, until coloured and tender.

If using the grill, preheat the grill first. Lightly oil the vegetables and grill over medium-high heat until tender, turning several times. Serve hot or warm.

To make the dip, put the yogourt in a fine mesh sieve lined with a clean kitchen cloth and position over a bowl. Let stand 30 to 45 minutes. Discard the whey. Add remaining ingredients to the drained yogourt. Mix well. Serve chilled.

Cook's note: Beets are the exception to this basic method of roasting root vegetables. Roast beets whole if small, or quartered if large, skin-on. Wrap handfuls in a double layer of aluminium foil, and cook in a hot oven until they give to a gentle squeeze, an hour or so. Peel them while they are still warm. Eat roasted beets naked or dressed in a small amount of your favourite vinaigrette.

Her former father-in-law's tree-planting resulted in the gift of a protected microclimate at Tam Anderson's farm near Bon Accord.

s is for squash

Prairie Gardens Adventure Farm Bon Accord
Tam Anderson

On a busy summer day, Tam Anderson waves as I thread my way through crowded aisles, past pallets stacked with fruit and vegetables in her on-farm store. I happily lose myself among the burgeoning squash plants and the smell of warm soil. Tam is briefly at my shoulder as I traverse the patio.

"It used to be strawberries and seniors, the generation oriented to picking, cooking and canning. But now they call and ask if we can pick them a pail. Our new demographic is young families with strollers," she tells me. "They want a place like Granny's that they can call their own and feel comfy coming to. They don't expect a working farm, and they're surprised when they see it is. But they all thank us for being growers, and doing what we do."

The farm experience creates a sense of community, and Tam and her staff answer a million Granny-like questions: "How is it grown? How do we cook it?"

"Safe, old-fashioned fun breaks down the barriers you might find in the city," she explains. Since Tam's former in-laws, John and Sheena Chedzoy, opened Prairie Gardens in 1956, over a million visitors have adopted this sunny spot north of Edmonton as their own. Tam bought the business when her marriage ended. "Farming is integral," she says. "My guests see authenticity when they see the soil and tractors and crops in the fields."

Tam is a pioneering woman in the farming and market gardening world. She was born into a family of English market gardeners; her aunt Anne founded Vale's Greenhouse in Black Diamond, and Tam studied horticulture at Olds College.

Pattypans, tender-skinned summer squash, have become a staple in Tam Anderson's market garden.

"I learned I loved being at the farmers' market, interacting with people," she says, "but I didn't enjoy the pressure of picking at 3:00 AM for fresh at 10:00 AM at the stall. So I started a u-pick."

When she noticed changes in local weather patterns—more dry weather and later frosts, later springs and warmer summers—Tam moved from strawberries to pumpkins and squash as her primary produce. In 2010, her staff harvested twenty-five thousand pounds of pumpkins, including orange, ghost-white, white with orange stripes and flecked like damask, temperamental tiny ornamentals and undemanding jack-o'-lantern giants.

"We start everything in the greenhouse, and transplant them to the fields, all by hand," Tam says. She rattles off the names of squashes, a flurry of words that sound like nursery rhymes: butternut, goldrush yellow, cousa, kabocha, acorn, spaghetti, pattypan, zucchini, hubbard blue, Lebanese. "I have a panic attack at the beginning of August each year, thinking there won't be any squashes or pumpkins, there's no fruit set yet. But in the last couple of weeks in August and September, they always put on size."

She and her staff always experiment at the grill. "People want to spend the day, and by the way, what's for lunch?" A grilled zucchini burger is perfect fodder for people who thought they'd seen all the ways to eat up their proliferation of zucchini.

Tam pins her success on the shoulders of those who came before her, and to the synchronicity of good geography. Her twenty-five-acre market garden, amid picturesque parkland, is situated on a deep trove of number one black topsoil. "It's prime growing land," she says gratefully. "We have a creek and a shelterbelt, a terrific microclimate thru the foresight of my former father-in-law, planting all those lovely trees in the late 1950s." Her husband, Terry Anderson, grows yellow and green peas, canola and grass seeds on another farm.

"Both my daughters, Laurel and Kate, have experiences they'll value forever: starting their own seeds, learning how to grow, harvest, cook." It all lends credence to Tam's belief in the unifying power of the family farm.

Tipi Creek Farm Morinville
Ron and Yolande Stark

"A lot of people living on acreages spend all their time on the highway driving kids to activities," Yolande Stark observes. "That's not a wise way to spend your kids' time when there is so much to do in the country."

When Yolande and her new husband, Ron, bought a u-pick strawberry market garden on sixty acres just off the Sturgeon River northwest of Edmonton in 1992, they decided to not make that mistake. Yolande had some business courses under her belt and Ron was an apprentice carpenter. "We were farm kids. We wanted to give our kids that freedom to roam and be kids without restrictions, with exposure to animals and how their food was grown."

Two years after they bought the farm, Ron and Yolande were approached by a group of people who had heard about Community Supported Agriculture and were looking for a grower. "The farm had beautiful soil. We weren't planning on growing large-scale, just for ourselves. But Ron just looked at the soil and said, 'wow, I want this, I can do something with this.' When the CSA idea was presented to us, it was worth a trial."

What became the province's first CSA started small. "We sweated a lot for twenty shares," Yolande recalls, laughing. Seed volumes and planting frequencies evolved by trial and error. Yolande chooses open-pollinated heritage seeds and applies a few simple but sensible rules that make it easier for her to count and divvy up produce. She plants in one hundred-foot rows; chooses small varieties of prolific squash, pumpkin, watermelon and cantaloupe, which thrive in the farm's black soil and tree-edged microclimate; she keeps subscriber numbers to a manageable number.

The magic number became forty families. "We tried as many as seventy," Yolande says, "but we lost the people aspect of it." When community and relationships are the driving force, losing track of faces in a busy blur is the biggest error to be made. "I can remember all the people who come, and their kids' names. Food is so fundamental in our lives. It should not be rushed."

Yolande and Ron do most of the work, with occasional help from various family members. A work schedule brings their subscribers to the farm several times annually.

"People want the farm experience, that connection. Some absolutely love coming out," Yolande says, laughing again. "It's a problem and it's not. Twenty-five percent of our subscribers would be here every week. On my part, it's that many more people to manage. It's a beautiful problem, isn't it?"

"Farmers' markets never appealed to either one of us," she observes. "We could have done very well—St. Albert's market was just starting when we began. But a CSA means we have a guaranteed market. We grow food the way Ron's grandmother did, as close to nature as possible without playing around with genetically modified seeds or chemicals. We aren't certified organic. We review it with our subscribers from time to time, but they know how we grow things."

That awareness is exactly what Ron and Yolande intended to cultivate back in 1992 when they started looking for rural Albertan land.

Linda's Market Gardens Ltd. Smoky Lake
Linda and Don Christensen

Linda Christensen's swinging metal gates into her farm near Smoky Lake are four feet high, not too big to obscure light but sturdy enough for privacy, and symbolic of the divide between work and home. "Anyone could drive in at 7:00 AM or 11:00 PM," she says a little sheepishly. "You want a family life, so the business opens and closes at certain times, morning and evening, with the gates."

In 1982, Linda, a young at-home mom with two small children, planted two one-hundred-foot rows of pickling cucumbers. "Don and I had moved back home, to the mixed farm where my parents raised me and my sisters. I was the youngest, and I'd done all the outside work as a kid with my dad," she explains. "I didn't want to go back to office work, but I needed income and communication with adults."

Linda put an ad in the local paper and sold her entire crop. A year later she planted a quarter-acre garden. When she couldn't sell all her produce at the gate, she went to the Smoky Lake farmers' market. Over bags of cucumbers and squash and corn, Linda heard customers say, "I've seen your sign on the highway, but the farm is too far out of the way." She and Don started looking for land on Highway 28.

Highway 28 is a through-route from Edmonton to St. Paul and the Cold Lake Air Weapons Range, and the gateway to the region's many lovely lakes. In 1998, Don and Linda found the perfect spot for a store alongside the highway just outside of Smoky Lake.

Thirty-five acres of market garden space and ten thousand square feet of greenhouses are split between the two sites. A divided land base minimizes the risk of being wiped out entirely by summer hail, and as Linda foresaw, the new site adds visibility. Each year, Don plants sunflowers in profusion along the highway's verge as an eye-brightening advertisement.

Two years after the new location opened, Don's employer, Alberta Agriculture, closed fifty offices throughout the province, and Don's job as a beef specialist vanished. "We could sell, and move to Stettler, where he could get work," Linda recalls thinking. "Or we could expand here." Don became the production manager of the sprawling fields filled with strawberries, the region's famous pumpkins and eight varieties of squash. Linda took on the staff and store, selling their straightforward produce. Some crops are directly inspired by their staff of Mexican labourers—cilantro, habanero and scotch bonnet peppers, and tomatillos that spread like weeds.

In early spring, the farm does a booming business in flowers and bedding plants. After Mother's Day, the highway location opens, and all retail

Not just food for birds, sunflowers provide a cheery farm marker on Highway 28 for shoppers looking for Linda's Market Garden.

business happens there, so Linda has a little peace behind the gate.

Smoky Lake, she says, "hasn't grown from when I was a girl, still a thousand people. The biggest thing is the pumpkin festival, and it only lasts one day. It's a nice place to raise a family; there's no pollution, no industry, but that's not a lot to bring people back."

Part of the challenge of life in a small town is a lack of local labour. Four Mexican labourers work in the fields and greenhouses each season. "Without the foreign worker program we would be out of business," Linda says. "Albertans don't want to do hard field labour. No one wants to do it, not since the oil boom in 2002. Men make more on the rigs. Nothing has changed."

The Garden of Van Ee-den Rosemary
Aaron and Barb Van Ee

When Barb Van Ee saw some of her customers picking vines and leaves off the plants in her u-pick squash and pumpkin patch, she wasn't dismayed. "Sometime the crops we plant are eaten in ways we don't know," she told herself philosophically. But when she saw squash blossoms in her customers' hands, she had to act. "The plants need their flowers in order to set fruit for picking," she explained to the Pakistani and Philippine women who were harvesting that day.

Many of the vegetables Barb plants in her small market garden near Rosemary are there because of the ethnic families who have taken root in nearby Brooks to work in the slaughterhouses. "Fish pepper, for instance, I grow it now; a tiny striped pepper, it's insanely hot, hotter than bird chiles," she says. "And eggplant, so beautiful to grow. I've introduced people around here to some different types of vegetables they might not know, all because my customers asked me to grow them."

As a young mom, Barb fed her kids from her own garden while they were growing up. She has an abiding passion for growing things, and is particularly fond of the myriad varieties of squash that she seeds. "Flying saucer squash," she says, and laughs. "You know the ones I mean, yellow and white, with the little scalloped edges?" She's referring to pattypan squash, but "flying saucer" is as fine a description as I've heard. "We grow pumpkins, tiny sugars and huge jack-o'-lanterns, and ghost white, and blue—they're Jarrahdale or Australian pumpkins, actually, the outer shell is a sickly grey, but they're lovely orange inside."

Barb and her husband, Aaron, who farms grain and oilseeds, live in irrigation country in southeast Alberta. Barb, now in her early fifties, had a career as an early childhood educator in Lethbridge, two hours' drive away. When she retired in 2006, moving to BC to plant an orchard simply wasn't an option, so she persuaded Aaron to give her a five-acre patch close to Highway 550 that his irrigation lines couldn't water. She hand-watered her first crop of strawberries, and diversified as her clients began to make requests. "Green tomatoes are a big hit with my Pakistani clients. I like Romas, and I've gotten everyone else talked into them too. Hot peppers like habanero, cayenne, jalapeno, and the sweet bells too—I grow all the makings for salsa."

Barb's learned to label her plants and rows with images to help multilingual visitors find what they want to pick. Many people who show up at the farm have no idea how potatoes grow, or how to dig them. "What's most shocking is people who don't know carrots. A guy came out from Calgary who thought carrots grew above the ground orange." She sighs. "People who have never done it just don't know."

Barb edged the patch with sunflowers and corn as an attention-getting windbreak, started attending the weekly Brooks market and began hiring local high school kids in the summer. She mixes a concoction of fish fertilizer and molasses to spray on her clay loam, and grows cantaloupes, starting them in the greenhouse, then covering them in the field with tunnels or "hot hats," a poly top hat over each plant to hold in heat.

She's been listening to comments at the market and at the farm. "People don't like to pay what food is worth, and they don't know how much it costs to grow here. My strawberries cost twice what California berries do

Squash blossoms are popular in Italian, Pakistani and Philippine cuisines.

in the supermarket. But ours are picked ripe and they're unsprayed, and the nutrition value from the field to the plate is way higher than food that's shipped. That whole cheap food thing hasn't gone away. Maybe you can change it with education, but people don't pay attention until it's personal."

roasted squash with peppers, corn, feta and smoked paprika

This mélange had its beginnings in the "three sisters," the classic Native American planting combo of corn, squash and beans, and succotash, the traditional blend of corn and lima beans that became a cheap and popular Depression-era standby. Squash's sweet and mild nature is always better when it's backlit by something salty—a little feta, for example—and something hot, if spicy is to your taste. Use the oven in winter, the grill in warmer months. Serves 4.

4 ears of corn or 2 cups (500 mL) corn kernels
1 onion, sliced
2 cups (500 mL) sliced winter or summer squash
1 red bell pepper cut in 1 in (2.5 cm) dice
1 jalapeno pepper (optional)
2–4 Tbsp (30–60 mL) extra virgin olive oil or cold-pressed organic canola oil
kosher salt and freshly cracked pepper to taste
½ tsp (2.5 mL) smoked or sweet paprika
½ cup (125 mL) crumbled feta cheese
2 Tbsp (30 mL) minced fresh parsley or chives
1–2 Tbsp (15–30 mL) extra virgin olive oil or cold-pressed canola oil,
    for drizzling

Preheat the oven to 400°F (200°C). Line a baking sheet with parchment paper. To cook corn on the grill, pull back the husks, remove all the silk, and replace the husks over the entire ear, then soak each ear in cold water for 20 minutes. Accomplish the same thing in the oven or in coals by wrapping shucked ears in foil. If using corn kernels, add them to the raw vegetable mixture prior to roasting.

Toss onion, squash and peppers in a bowl with a generous drizzle of oil. Season with salt, pepper and paprika, then spread the vegetables in a thin layer on the baking sheet. Roast uncovered for 25 to 30 minutes, stirring several times, until the vegetables are tender and beginning to show some brown edges. If grilling, place directly on the grill bars on medium-high heat without using a basket (a basket generates steam) and turn several times, cooking until tender and slightly charred. Cut the corn kernels from the cobs and toss in a bowl with the roasted vegetables. Mix gently with the feta and herbs. Add a drizzle of oil. Serve hot or warm.

Juicy ripe tomatoes and toasted walnuts combine to make
the perfect summer sandwich of bruschetta on grilled bread.

t is for tomatoes

Hotchkiss Herbs & Produce Rocky View
Paul and Tracy Hotchkiss

People love or hate tomatoes. Paul Hotchkiss loves 'em. In the late '80s, after making a tidy fortune in the oilpatch, he planted heirloom seeds in a cobbled-together greenhouse, hoping to grow delicious tomatoes for his favourite lunch of BLTs. The plants grew, triffid-like, taking over his life, until he realized that he was growing sufficient amounts to sell instead of simply giving away his surplus. At peak, over one hundred heirloom varieties—Brandywines in several hues, purple-bruised Black Krims, citrusy Green Zebras, Tigrella, Cherokee Purple, Pineapples with glowing lines of red through orange flesh—grew in symbiotic splendour with tiny hopfrogs, ladybugs and other friendly life forms. Paul had plenty of healthy, delicious certified organic tomatoes until tobacco mosaic virus was accidentally introduced, imported on a plant from southern Ontario.

Eradication efforts included old wives' tales, like washing hands in milk, and pure science. Staffers changed clothes at every row-end and rinsed their hands in bleach solutions. Ultimately, Paul used a Bobcat to empty and refill two greenhouses with over fourteen hundred cubic yards of soil.

The new soil didn't help. The virus returned, and Paul immersed himself in studying plant genetics, breeding multiple generations of tomatoes, aiming to produce disease-resistant tomatoes that he could cross with his susceptible heirloom varieties.

"Would I have started the tomato thing if I knew the virus would come?" Paul rumples his red hedgehog hair and rubs a bare knee. "I enjoy it, despite the hellacious hours. I work at home, I can see my daughter, Olivia, and my wife, Tracy. We have a pair of pugs, horses, hay to bring in. I value my own personal freedom, I can take ten minutes off and throw a fly in the pond and catch a twelve-pound rainbow." Not that he does very often. Paul has been a fisherman since age four, and remembers going up Highway 93 to fish in pothole lakes with his dad, Harley, the hockey magnate. "We were out in a rowboat, three kids, each with two hooks, catching perch two at a time. We kept Dad busier than a one-armed wallpaper hanger."

Tenacity was vital to what became a classic Mendelian genetics study. "What I've done is not rocket science. It was a function of observation, data collection and perseverance. There's nothing loopy, no gene splicing; we went right back to counting the peas. It was a lot of fun, if you have a particular mindset. But those classic plant-breeding people have all been replaced by gene splicers, and I was left talking to old scientists and seed-savers sitting on back porches across small-town North America."

Manually cross-pollinating plants is a finicky task involving precise timing and a microscope, a gift from Tracy, who cheerfully admits to nei-ther a green thumb nor a taste for vegetables. To learn how, Paul watched bumblebees at work. "It was an interesting challenge," he says. "There was no path, so you don't know if you are veering off it or not."

Life on an organic farm is complicated by Paul's own nature. "If a little compost is good, I thought more would be better," he admits. It wasn't. After rigorous soil tests, modifying one variant at a time, Paul got back to where he started, using field soil mixed with peat and negligible compost to create the ultimate soil nutrient balance.

The virus paradox illustrates some of the reasons we don't find big boxes of heirloom yellow Brandywines for sale at supermarkets. Paul's current colourful tomatoes are a new generation of his in-house cross-breeding program that successfully grafted heirlooms onto stabilized home-grown root stock.

Paul, Tracy and their Mexican staff grow speckled romaine, aru-gula, rainbow Swiss chard, Purple Haze and White Satin carrots, Mediterranean cucumbers, Nor'easter pole beans, yellow cauliflower, pea and sunflower shoots, and microgreens on ten acres of fields and in two acres of greenhouses, with ten thousand feet added in 2011. Heirloom vegetables lovely enough to paint are the Hotchkiss trademark, served and sold in restaurants and shops in Calgary and the Bow Valley. They make growing a tomato look easy.

Paradise Hill Farm Nanton
Tony and Karen Legault

The late spring sun reflects across the ox-bow bends of Mosquito Creek as I drive east from Nanton in 2004. Tony Legault, limping slightly, steps stiffly over the string that stretches through empty air between two poles. "Welcome to the new greenhouse," he says, a cheerful robin of a man. He shakes my hand, eye-to-eye. "Care for a tomato?"

Four years previously, on a warm summer evening in 2000, Tony had been just as sanguine as he looked at his wife, Karen, and said, "We have seven thousand pounds of ripe tomatoes to pick." It was their first harvest.

"It seemed like a lot," Karen recalls. "We were pretty scared." Tony was recovering from a bad auto accident. Karen was a stay-at-home mom.

The tomato caper had a low-key start in Vancouver in 1988. Karen and Tony were suffocating in the rat race, Tony as a mechanic, while Karen, a farm-raised 4-H girl, worked in the agriculture offices of the Pacific National Exhibition. A family hunting trip near DeWinton convinced them to relocate to Turner Valley. Karen found work with the Calgary Stampede, and in 2000, kids in tow, they moved to Nanton and Karen quit her job. Her father, Gordon Souter, and her city-boy husband were discussing options when Tony said, "I wish I could be a farmer."

Her dad, retired and a BC resident, said simply, "Why not?"

"Tony was taught to get a trade and hang on for dear life," Karen recollects. Her husband wasn't a risk-taker. But Gordon's words helped the young couple change their focus. "Instead of looking for reasons why not," Karen says, "we started to look for reasons why we could." They discarded possibilities that required deep monetary reserves: the cost of quota ruled out dairy and chickens; a quarter million bucks for a used combine eliminated grain as well.

Sheep seemed likeliest until Gordon, aware of BC's burgeoning greenhouse industry, piped up again: "Have you thought about greenhouses?"

A little more parental nudging helped them locate an existing research greenhouse just east of Nanton, along Mosquito Creek, close to the heavy traffic and potential consumers coursing along Highway 2. "The place was filled with junk, and I couldn't see the possibilities," Karen says, "but I trusted my husband; he's a visionary."

In 2004, Tony Legault cleared land beside his family's greenhouse to expand what has become a thriving pesticide-free tomato business.

When they decided to produce a specialized crop to sell to local grocery stores instead of at seasonal farmers' markets, the couple chose hydroponic tomatoes. "How you grow a tomato makes a big difference in its taste," Karen explains. "At the grocery store I can find decent cukes and lettuce, but not a tomato that tastes good."

Fast-forward to those seven thousand pounds waiting to be harvested. Tony bought a dehydrator and dried tomatoes, one hundred pounds at a time. They made salsa too, and sold dried tomato chips, jars of salsa and fresh tomatoes at farmgate while Karen talked to local produce buyers.

One stop was the Deer Valley Co-op, in the south end of Calgary, where the produce manager said, "I think we could use a good local pesticide-free tomato." The next day, Karen loaded tomatoes into the station wagon and took them in.

"People just needed to try them," she says, giving full credit to the literal power of word of mouth. "The produce guys at three stores liked them and suggested them to their customers."

The Legaults celebrated their tenth "tomato" anniversary in 2010, and now supply twenty-two of twenty-three Co-op stores in Calgary and Airdrie with local pesticide-free tomatoes from March to November. "We don't have a broker. We are direct marketers," Karen says proudly. "We are the truck drivers, the growers, the sellers."

Hillside Greenhouses Bowden
Carmen and José Fuentes

Carmen Fuentes looks like an underage child in an oversize red T-shirt behind the market table, the Innisfail Growers sign towering above her. The thirtysomething mother simultaneously makes change and holds out a basket of neon-orange tomatoes. "Try these," she says, her smooth face opening in a slightly crooked grin. The woman before her bites a tomato and nods, her mouth full as she puts cash in Carmen's brown hand.

Carmen started growing things in greenhouses when she was fifteen. Bored, her high school friends away in Mexico one Easter, she bought tomato seeds at a local nursery. After she successfully grew a crop in the garage, Carmen's dad built her a garden-style greenhouse. It was love at first bite.

Synchronicity took over: Dad, an electrician, knew Shelley Bradshaw, a member of the Innisfail Growers marketing co-op. While wiring Shelley's carrot storage shed, the topic of tomatoes came up, and he came home with an invitation for his daughter. "I had work waiting, really," Carmen recalls of completing a greenhouse management program in Olds College's horticulture program. She grew tomatoes for Innisfail Growers for two years, then took a year off in 2000, waitressing and saving her tips. "When people find out a waitress has a goal, if you do a good job and have a good story, they can be really supportive." Carmen's tip jar collected enough silver to put up a thirty-by-fifty-foot greenhouse. A second stint of double jobs—opening at a coffee shop and closing at a pizza joint—funded her greenhouse's doubling to three thousand square feet.

When she returned to the Innisfail Growers' fold, Carmen grew orange Bolzano, grape and cherry tomatoes in pots, filled with a mixture of soil and peat moss. "You get more flavour from soil, even if it seems more old-fashioned," she says, almost dancing behind the market table as she hands me a cluster of tomatoes.

At the pizza joint, Carmen met José Fuentes, a Chilean on a horti-culture exchange who was washing dishes and attending Olds College

Carmen Fuentes strives to balance work and family in her tomato-growing career. She takes her lead from Leona Staples, co-owner of The Jungle, who has raised three sons as well as countless crops of fruit and vegetables on her family farm.

as an agriculture business student. They married in 2003. By then, Carmen had an established niche with Innisfail Growers, and José had a job in Edmonton. "It was a big decision. One of us had to give up their job, so José started apprenticing as an electrician and working in the greenhouse."

Baby Benjamin arrived in 2005, and in 2009, they expanded the greenhouse to sixteen thousand square feet. José continues to work as an electrician. "We could scrape by without that income," Carmen says, "but we'd need a good twenty years of hard work." After the first harvest of twenty-five-thousand pounds of produce, Carmen was close to burnout, so they brought in seasonal Mexican labourers.

They close in November, rest in December, plant in January. From February on, Carmen and two employees do the same thing every day—pick and pack tomatoes to sell at the many markets the Innisfail Growers frequent.

The expanded greenhouse is situated on a bare acreage, and Carmen is saving again, but this time, it's for funds to build a house and workshop. "We want to attach a building to the greenhouse, live upstairs, and use the main floor as a dual-purpose space. Eventually I want a little farmgate store with a kitchen, but you need to be living there if you have people stopping by to shop. We have tourists going by on our road. I want them to stop here." Carmen makes salsa, crabapple jelly and poppycock (caramel popcorn and nuts), all of which have become staples at Innisfail Growers' booths.

Carmen was twenty-three when she started attending the Innisfail Growers' group meetings and saw firsthand the organizational skills that bankrolled its success. "I found all the members intimidating, very successful, all very intelligent, but there's that twenty-year age gap!" Leona

Staples, the tireless co-owner of The Jungle, is especially inspiring to Carmen. "Leona still has kids in school. I keep wondering how she's done it all with three boys. We've never had to put Ben in daycare. That was always one of my goals as a business owner. Life is good. I've had bosses, and it's nice to not have one. I'm grateful for the Albertans who shop at farmers' markets and pay for local food. We have loyal shoppers. Without them we wouldn't be here."

Tomatoes are more flavourful when grown in soil, says Carmen Fuentes.

tomato, walnut and cilantro bruschetta

Based loosely on *muhammara*, a classic Turkish relish, this spread is spectacular on simple grilled bread as a lunch or an appetizer. It works equally well as a sauce for grilled or roasted fish and chicken. In corn season, add a handful of grilled or roasted corn kernels to the mixture. Yum. Serves 4.

1 cup (250 mL) fresh walnut halves
2 cups (500 mL) diced ripe tomato
2 Tbsp (30 mL) pomegranate molasses
1 tsp (5 mL) toasted and ground cumin
½ tsp (2.5 mL) sumac (optional)
juice of 1 lemon
1 garlic clove, minced
¼ cup (60 mL) olive oil or cold-pressed organic canola oil
1 Tbsp (15 mL) walnut oil
2 Tbsp (30 mL) minced fresh cilantro
kosher salt and hot chile flakes to taste
crusty sliced bread for the grill
1–2 Tbsp (15–30 mL) olive oil or cold-pressed organic canola oil,
      for drizzling

Preheat the grill or oven to 350°F (180°C). Put walnuts on a baking sheet in a shallow layer and toast them for 10 minutes. Remove from the oven and cool thoroughly, then chop with a knife. Set aside. Stir together tomato and pomegranate molasses in a bowl, then mix in cumin, sumac, lemon juice and garlic. Add the oils, whisking well. Stir in cilantro, toasted walnuts, salt and hot chile flakes. Set aside. Drizzle sliced bread with oil. Grill or broil. Remove from heat, garnish with sauce and serve immediately.

u is for u-pick

Sprout Farms Apple Orchards Bon Accord
Amanda Chedzoy

Harvested from well-known valleys—BC's Okanagan and Similkameen; Ontario's Twenty Valley, sheltered by the Niagara Escarpment; Nova Scotia's Annapolis—apples could be regarded as Canada's national fruit. On the high plains, berries are widely considered the natural fruit. Apples do grow in Alberta, although as Edmonton-area orchardist Amanda Chedzoy volunteers, they're chancy. "It's the unpredictable weather; you just never know if you will have a crop. But I like the challenge." The chanciness is why Amanda has a related job, as a self-employed certified arborist.

Amanda is in her late fifties, with well-muscled arms and hands. She and her husband, Tom Carleton, live a physical life on twenty-eight acres near Bon Accord, on land that has grown trees since her childhood, when her father managed the Alberta Tree Nursery. "It was a cool place to grow up," she says. "It turned me into a country girl."

As a young woman, the "country girl" worked in New Zealand for two years, tending sheep, picking apples, and weighing and testing wools in a lab. When she returned to Canada and married, she didn't want anything to do with agriculture. But the trees had their roots in her, and it was a tree-moving business that Amanda and her first husband founded in Edmonton. In 1975, with young kids, they bought the nursery.

After they divorced, Amanda spent the summer of 2004 planting three hundred saplings, mostly apple, with an eye to selling u-pick fruit when the trees matured. She has seven hundred trees now, about half of them bearing fruit. "Apples were the easiest to grow," Amanda says. "I thought a u-pick would be fun." She was glad to leave behind the relentlessly physical work of the nursery, although people still call hoping to buy trees.

In the orchard, there are dozens of apple cultivars, often more than one

Apples in Alberta? You bet! In Brosseau and Bon Accord, apples have thrived since scientist Cecil Patterson began his plant research into hardy varieties at the University of Saskatchewan in 1922.

on the same root stock. "We grafted on old prairie-hardy varieties when Doctor Patterson was at the university." She means horticulture pioneer Cecil Frederick Patterson, who initiated fruit breeding at the University of Saskatchewan in 1922. In his thirty-eight years of breeding research and growing, he developed Russian-derived apple cultivars like McLean, which Amanda describes as a small, juicy, sharply flavoured apple with pink juices. Another variety, Harcourt, which originated in Gleichen, tastes somewhere between a McIntosh and a Gala, tangy but soft-textured.

A two-hundred-tree demonstration orchard planted in 1980, including other fruits as well as apples, has grown over the years, with more multiple-grafted cultivars. "We started this orchard when we sold trees," Amanda explains, "because we wanted people to taste the apples and see the mature trees." When the weather allows a robust u-pick crop, Amanda gives her customers a taste of all the apple varieties she grows.

Sweet yellow Brook Gold plums are what Amanda loves best. "We have ten varieties of plums in that orchard. Some fruit every year, some only fruit every ten years. The thing is, they bloom really early, and we often get frost when the blossoms are breaking. I don't reckon on them being a big part of the crop, they're just a bonus."

The pears are small, five centimetres long. "People seem willing to deal with the smallness," she says. "They are genuinely thrilled to buy a local product, and it's such a treat to eat them fresh." Windfalls are eaten by the sheep that serve as grass mowers, an ecological way to minimize any risk of bug infestation and keep the wasps down.

Amanda found an old wooden screw press, and is looking into making cider. Not for sale, not at this point, she says—cider is an unknown quantity to the Alberta government, and there are no regulations governing its production as yet. So that too is a chancy matter. But Amanda isn't troubled. "We are small-scale; we don't want thousands of people running around the farm. School tours and a few u-pick families at a time allow us time to visit, show the sheep to the kids, and give them a farm experience. It's the pace I like."

Antelope Creek Road Berry Farm Brooks
David and Elizabeth Houseman

As a seasoned hunter friend likes to remind me, meat is a crop that we harvest. For most of us, that means going to a store and staring at impermeable cryovaced packages, or to a butcher where we watch muscled hands cut two-inch rib-eye steaks. Rarely do we see dinner on the hoof. The occasional time we do see a free-ranging animal, we don't connect it with the steak on our plate. That says a lot about the distance we have travelled from the land, literally and metaphorically. But there are consumers who have a closer relationship with the animals they eat, and with the people who raise them.

David and Elizabeth Houseman of Brooks have a prolific u-pick berry patch and orchard—two hundred and fifty feet of cherry trees, a dozen rows of raspberries, blackcurrants and thousands of strawberry plants. In 2010, their clients and their staff of Hutterite women picked over four thousand pounds of raspberries and saskatoons. "I don't have to say the word organic," she says. "We aren't certified, but people know there are no chemicals on their food and are more willing to buy it. They want to know who grew it and who you are."

Sun-ripened berries are best for simple desserts.

Elizabeth spends hours each summer making jams, jellies and preserves that she sells at the farmgate and at the market. But the fruit is only part of the farm's business. Elizabeth and David's u-pick berry farm is also a u-harvest meat destination.

"The majority of our pastured animals are sold live," Elizabeth says. "Many of our regular customers are ethnic, they have moved from places like Africa, the Philippines and Mexico to work in the slaughterhouses in Brooks. They like their meat fresh. So they take our birds and the occasional lamb home to butcher themselves."

About two hundred heritage turkeys, chickens and ducks are pastured in the fair-weather months at Antelope Creek Farm. Broad-breasted Bronze turkeys destined for holiday platters are outlived by Muscovy ducks and Araucana chickens, gawky white-tufted Chilean birds that are also called South American rumpless and are one of two breeds that lay blue eggs. None fly away, although they could in theory. At worst, the turkeys regularly perch on top of the barn, and the ducks swoop overhead. Elizabeth often puts turkey eggs under the ducks for hatching to prevent the clumsier turkeys from kicking their eggs around and breaking them.

The farm had its beginnings when the couple met at David's previous market garden. He was divorced, with seven children; Elizabeth, a widow, was raising three kids. She went to his raspberry garden to pick berries to make into wine for her eldest daughter's wedding. He wrote a note that the busy mother of the bride tucked into her purse unread. "When I told my daughter," Elizabeth recounts, "she said, 'what did it say, Mom? Read it, read it!' Well, he wanted to have dinner with me."

Three years after Elizabeth and David's marriage in 1997, the newly blended family looked for a patch of ground with access to water, the farm that David, a born farmer, had always wanted. "We found a quarter, bare hills near Antelope Creek, and started planting. David said, 'we'll plant three thousand strawberry plants,' and we did, but then he said, 'no, I meant three thousand of each kind!' I was ready to drop dead of exhaustion," Elizabeth recalls.

David kept his off-farm job as a province-wide Saskatoon berry harvester for five more years. He's in his mid-seventies now, a fit and limber man who farms full-time; his wife says they'll farm for as long as he's able. None of the blended family's kids shows any interest in farming. Elizabeth sounds regretful as she adds, "The country needs younger people getting into this industry. There's too much of this 'don't know your neighbour' in the cities." What's apparent as another animal leaves her yard is that the Housemans' regular customers know David and Elizabeth well, and appreciate the chance to harvest their own food.

The Blooming Fields Didsbury
Mary-Ann and Pim van Oeveren

I finally notice the sign in a wooden frame positioned on the north side of The Blooming Fields, on the café's exterior wall, angled above a seven-foot window. *Your local food.* How did I miss that? The answer is obvious when I walk around the welcoming café and simply stand still. The place is beautiful, as lovely as any parkland in Europe. The land gently slopes west from the stone patio to the lovely garden, and flattens out at the orchard. Wooden signposts on tidy walkways indicate the species and variety of plants. South of the café, more fields and beds stretch, more tidy rows and patches.

"We planted all the friendly herbs—parsley, lovage and garlic, oregano, thyme, summer savoury—whatever wants to grow here," Mary-Ann van Oeveren says. A slight Dutch accent candies her words. "Everything we thought there would be a market for: vegetables, spuds, fruit. We wanted our own garden; we planted a bit bigger so we could sell the extra. It was that simple."

Mary-Ann and Pim emigrated from the Netherlands to Mission, BC, in 1994. Pim, then a thirty-nine-year-old landscaper and nursery grower, thought the move would be an easy transition. Their business thrived, but to Mary-Ann, the deeply populated Lower Mainland was never home. "It felt like one big city, and we were regarded as immigrants in a city of transients. We wanted a home." They followed their adult children to a

The wooden stairs at The Blooming Fields lead to carefully tended rows of fruits and vegetables.

piece of land near Didsbury in 2003. Mary-Ann recalls that rural Alberta felt remote and climatically distant. "It was like the middle of nowhere, and I questioned our sanity about the cold weather."

Anxious to work with people, Mary-Ann taught floral design at Olds College while Pim built a nursery, garden beds and an addition to their home to serve as a shop and café. By 2005, when the blue and white spruces were established and the saskatoon bushes were fruit-laden, the couple was on the cusp of fifty. Produce was limited when they opened the doors to the café, u-pick garden, orchard and nursery. "We had whatever was ready, and no greenhouse at the front yet," Mary-Ann remembers. Since then, the orchard has expanded to include tidy rows of raspberries, black and red currants, a clutch of rhubarb, but no strawberries. "We had such bad luck; every year they were too late or [there were] so few berries," Mary-Ann mourns. "Myself, I lean to saskatoons and blackcurrants, as berries and for the jams I make and sell." In 2009, the cherries they planted four years previous began to bear fruit—Evans and University of Saskatchewan varieties, Romeo, Juliet and Carmine Jewel.

Mary-Ann is busier than she had expected, tending her shop and country-style café, which is filled with luxurious indulgences for women. Food-related gifts and teapots, she says, "things ladies want for themselves or their best friend, pampering knick-knacks, a little luxury in the country."

When I've finished picking, I sit on the sunny patio, among the visitors whose bags and bellies are blooming with plenty. The sign mounted on the wall reminds me that the devil is in the details. *Your local food.* On a map, as with a camera lens, details narrow the focus until I see a particular plant on a particular patch of land. That's local. That's precisely what blooming fields and gardens offer to observant eaters.

Chef Scott Pohorelic, left, now a culinary educator at SAIT Polytechnic in Calgary, shown with Broxbum co-owner Paul de Jonge, recalls using Broxburn strawberries as late as November on his seasonal menus at Calgary's River Café.

berries in yogourt cream with green pepper sauce

This dish relies on the serendipity of cool, unctuous yogourt and perfectly ripe fruit. Use fresh fruits and berries in summer; in winter, try poached pears and rehydrated dried apricots or cherries. Serves 4.

yogourt cream:
3 cups (750 mL) yogourt (choose a gelatin-free brand)
2 Tbsp (30 mL) honey or maple syrup
½ tsp (2.5 mL) grated orange zest
½ tsp (2.5 mL) grated fresh nutmeg or cinnamon
fruit:
2 cups (500 mL) fresh strawberries or other ripe fruit
sauce:
2 Tbsp (30 mL) maple syrup or honey
1 Tbsp (15 mL) butter
1 tsp (5 mL) cracked green peppercorns (dried, not brined)
1 Tbsp (15 mL) orange liqueur
3–4 oz (90–125 mL) orange juice
½ cup (125 mL) whipping cream

Drain yogourt for 45 minutes through a fine sieve lined with a damp kitchen towel. Discard the whey. Add honey or maple syrup, zest and nutmeg or cinnamon. Spoon yogourt into wine glasses. Slice strawberries into halves or quarters. Divide among the glasses. Set aside while you make the sauce.

Heat maple syrup or honey and butter in a small sauté pan, bring to a boil, add green pepper, orange liqueur and orange juice. Reduce by half, boiling to evaporate, then add cream and return to the boil, reducing by half again. Divide the hot liquid evenly overtop the fruit in the wine glasses. Serve hot or warm.

v is for vegetables

Blue Mountain Biodynamic Farms Carstairs
Kristian Vester and Tamara Brunt

Kris Vester is most himself in a field. It's the place where his politics fall naturally into alignment with his life, where his burly frame is most relaxed, where his brown hands fit most comfortably. Even his words flow more concisely. His strongest fervour is for the well-being of the soil, and in the field, his natural shyness falls away.

Paying close attention to how things work has taught Kris that sustainable living is the only option. "I thought I was a leftist when I was younger, but I have learned how fundamentally conservative I am," Kris tells me, laughing. He's being ironic, and doesn't mean "conservative" as most Albertans construe the word, but rather, to imply the preservation and protection of all life and culture. "A true conservative pays more attention to natural capital than to money, to preserving the ecosphere above preserving the economy."

Kris, his wife, Tamara Brunt, and his preteen son, Niko, live near Carstairs on Blue Mountain Biodynamic Farms, which was founded by Kris' family in 1977. In his late thirties, this first-generation Canadian of Danish-German extraction is a well-read man with deep faith in God and the natural state of the universe.

Kris didn't plan to be a farmer, but a university education in Germanic and classical studies hasn't impeded his faith in farming. "At the time," he says, rubbing his big hands over his bushy beard, "I was frightened into believing I needed a degree to make something of myself." He has grown into an ambassador's role for his chosen field. As Slow Food Calgary's current convivium leader, Kris's populist style can ignite an audience. In a speech delivered in Calgary in 2009, Kris said, "The most important and radical work a young person could undertake on this earth would

Kris Vester used to sell his biodynamic produce and grains at open street markets like Calgary's Green Market 17, but is convinced that CSAs offer farmers and consumers the best opportunities for interaction.

be to grow food for local consumers, in a fashion which would reassert the farmer's dignified and essential role." He's a fervent believer in the resilience, resourcefulness and abilities of small farmers, who "are the most knowledgeable and adaptable producers of food on this planet . . ."

Kris and Tamara follow the biodynamic practices of Rudolph Steiner's holistic approach to growing. They limit the volumes they produce to what the two of them can do themselves with limited help: WWOOFers and SOIL (Stewards Of Irreplaceable Land) apprentices come for two-week stints. "Canadian kids don't want to work that hard," Kris comments. "They have a fair expectation of a life of ease and convenience. But it doesn't feel right to separate other people from their families to work here."

A wide range of vegetables and fruits—herbs, greens, root crops, plus laying hens—fill five acres of market garden tucked inside a quarter section. Their produce, and house-milled triticale, rye, oats and wheat, go to CSA subscribers, with the overage sold at Sunnyside Market in Calgary and through several local-food-delivery programs. In Kris' view, most grocery stores value profits over relationships and the greater good. The CSA structure "allows the consumers to share the risk with the farmer, who usually is the only one watching the skies darken."

A cultural change is coming, he says, when consumers will pay attention to more than the apparent price of a Big Mac. The real price is invisible: ecological and health costs that are being banked for payment by future generations. "I hope Niko will farm too. It's our insurance policy: we can always feed ourselves," he says, then warns, "Do not feel self-satisfied with your quality of diet or quality of life. It's about all of us, across all social and economic strata. We need to ask, ultimately, whether money is more important than food."

Nature's Way Veggie Patch Peace River
Lisa, Peter and Mary Lundgard

The acorn has landed close to the tree. Lisa Lundgard grew up with her three siblings and parents, Peter and Mary Lundgard, on the family farm north of Fairview, west of Peace River. In 2005, the family moved to a section of land near Grimshaw, just north of the picturesque Shaftesbury ferry across the Peace River. It's good farmland, with as much heat as Taber far to the south.

Peter and Mary follow holistic management principles in their personal and professional lives. Pastured Berkshire pigs, poultry, hair sheep, Galloway cattle, and Peter's pet project, leaf cutter bees that pollinate seed alfalfa, are all reflections of their belief in the soil and the grass it produces as the main cause and effect in growing good food. They went to the USA and New Zealand to study soil management, primarily learning to read and balance mineral content. "Compost is a good way," Peter says, "and compost teas, and microbial inoculants that can be broadcast or mixed into water."

Their second youngest, Lisa, feels a particular attachment to the land and subscribes to her parents' approach. "My biggest chore as a kid was moving the chickens every day. The chickens' pens are 'portable,'

Biodynamics
Rudolph Steiner, an Austrian philosopher born in 1861, founded the Waldorf school movement, which now has over six hundred schools around the world. He gardened too, and in 1924, devised biodynamics, believed to be the modern world's oldest non-chemical agricultural approach. One of the main principles of biodynamic farming is that the farm, like a human, is treated as a whole, a self-sustaining organism. Steiner advocated companion planting, crop rotation, and sowing and harvesting according to the lunar calendar. He also designed esoterica such as the use of minute applications of homeopathic-style infusions of mineral, plant or manure extracts to heal and revitalize the soil. Biodynamics has a wide following globally, with participating groups and members mostly concentrated in Europe. A trademark, Demeter, named for the ancient goddess of the earth, is used to indicate certified farms where biodynamic principles are followed.

but there are twenty of them. They move via sheer human strength," she remembers, laughing. "My brother did the pulling and I did the pushing."

Lisa, who has a psychology degree under her belt, is now in her early twenties. She's a lean, tall woman, her ponytail usually confined in a messy bun under her straw hat. "I like farm life, and space, clean fresh air, animals I have relationships with, and fresh food whose origins I know," she says candidly. Lisa started hearing about Community Supported Agriculture in 2006 and spent a day with Yolande Stark at Tipi Creek CSA, talking shop. "It sounded like the most feasible type of farming and business, to know you have a customer base. It's hard growing food, if something goes wrong without insurance or deep pockets to back you up, you lose everything. A CSA adds security and shares the risk."

Peter Lundgard is proud of his daughter Lisa's foray into farming with her CSA, located on the family's Peace Country farm.

Acquiring land, the biggest challenge for any young farmer, was no problem. Lisa tilled a half-acre plot on the family farm in spring 2010. Her partner in that first year, Sarina Piercy, had been her family's farming intern, earning her bed and board. They started tomatoes and cucumbers in three greenhouses, and field-seeded a garden of tough and stalwart vegetables, especially her favourite carrots, and lettuces she treated like "cut and come again" mesclun, (cutting the leaves and waiting for the plants to grow again). "Open pollinated seeds were the key, so I can save seeds for later years." In their first season, they grew for fifteen clients. It wasn't a lot of money, but Lisa decided to persevere on her own the next year.

The biggest challenge Lisa faces is labour. She's hoping to attract WWOOFers. "Lots of weeding! I want to find more efficient ways of harvesting, washing and bagging. Ideally, a CSA has loyal customers who want to weed and pick." She puts out a weekly newsletter that chats about the week's biggest crop—a jungle of tomatoes, or a flood of peas. "I love being able to tell the stories, and the relationships we create are amazing, taking the boxes to Peace River for pickup. It's so neat to create friendships."

Like her parents, Lisa is looking to the long view. She doesn't expect to be a CSA farmer all her life, but she expects to raise her family rurally when that day arrives, on a little piece of land with a milk cow and chickens. In the meantime, her garden is not just about the work. It's about the community that coalesces about the work and its fruitful results.

Holistic Management Training

Acres U.S.A. claims to be North America's oldest, largest magazine covering commercial-scale organic and sustainable farming, or eco-agriculture. Founded in 1975 by Charles Walters, the magazine offers field days, training seminars and workshops at leading American farms. One such seminar is on holistic management, which Peter and Mary Lundgard attended. As a model to making better decisions in life and in business, Peter says holistic management begins by defining who is involved, and writing a clear common goal in three parts: the quality of life or values shared by the group; the forms of production that will move toward that life; and a resource-based description of the production process.

Participants ask seven questions contemplating action. What effect will this have on the ecosystem? Are you dealing with a root cause or a symptom? Are you dealing with the weak link? What will be the biggest bang for the money spent? What is the gross margin analysis? What is the energy wealth source and use, specifically—where is the cash coming from to do this, and where is the energy coming from to do this? What effect does this decision have on our community or society?

What follows is an assessment of available tools: human creativity, money, rest, stock density, animal impact. Financial and biological plans and frequent monitoring conclude the process.

A high percentage of Newfoundlanders working in the oil city of Fort McMurray has meant a cabbage boom for Dunvegan Gardens. Cabbage is a staple in jig's dinner, the Newfoundland culinary classic of boiled salt beef and humble vegetables.

Dunvegan Gardens Dunvegan, Edmonton and Fort McMurray
Brad, Ron and Pauline Friesen

"Vegetable growing doesn't have the money power that landscaping does," observes market gardener Ron Friesen. He and his wife, Pauline, co-own Dunvegan Gardens. Ron's father, Bill, began the business in 1952, at Dunvegan, where the confluence of highway with river valley created a busy hub in the Peace Country. It's a stunningly beautiful site in the lea of the escarpment, adjacent to the suspension bridge across the Peace River.

Ron's son Brad operates an arm of the business in the burgeoning oil sands city of Fort McMurray—greenhouses, a strawberry field and about thirty acres in vegetables—although both agree that the commercial landscaping wing of Dunvegan Gardens pays the freight. "Food isn't it at this point," Ron says acerbically. "Consumers still aren't paying what it's worth in Canada."

It's a sad commentary. If what we spend our money on is a true indicator of what we value, then houses and Hummers are outscoring honeydew melons and horseradish.

It wasn't always so, Ron says. "My folks were substantial growers. Dad made a good living at it fifty-five years ago. When I was eleven, it was cucumbers and cabbage and corn and cauliflower. People took it home and froze it. Demographically and culturally, people eat differently now."

Supply systems have changed too, Ron says. "Nowadays, if you want to sell a cabbage to a Safeway store in Grande Prairie, it goes to Edmonton and then it comes back. Back then, we kept thousands of laying hens as well as the garden. We had thirty stops in Dawson Creek, it was bigger than Grande Prairie is now, and was a strategic point for shipping up the Alaska Highway. From Peace River, food was delivered up the Mackenzie Highway. Both had a vital part in keeping the north fed."

Ron and his two brothers bought out the business from their parents in the late 1960s. They slowly got pushed out of the Edmonton area, and food's slide to a lesser role in the business began. In 2011, garden centres and bedding plants generate the majority of the business in Dunvegan and Edmonton, out-producing patches of corn, vegetables and strawberries.

Eleven Mexican labourers work in the three locations. "If we want cheap food, we don't want to abuse the foreign workers who work the fields," Ron says sharply. "We have the cheapest food in the world in Canada. The Americans have freeloaded off the backs of Mexicans for years, and we don't. Our guys get a raise every year and they keep coming back."

Brad says, "We've been losing money to grow fresh produce here in Fort McMurray, but we're the only farm around. All around is muskeg and trees. Edmonton or Wandering River, where Highway 55 and Highway 63 meet, that's where farmland starts." Eighty per cent of the berries his workers pick from ten thousand strawberry plants are sold at the farmgate store, he says. "All the food we don't grow is trucked in. Yep, it costs more to eat here. But it's relative to wages—you can be making one hundred thousand dollars here and still not afford your own home."

Fort McMurray's population is ninety-five thousand and climbing, plus twenty-five thousand workers in camps within a one-hundred-and-twenty-kilometre radius. "We grow a lot of turnips and cabbage in our vegetable fields for the Newfoundlanders; there's a lot of them," Brad says. "They put 'em in their jig's dinner. Plus there's a huge ethnic workforce here, from Somalia, India, China, Mexico, Venezuela, Chile."

David Jensen grows corn in Taber, one of the few Canadian regions where the soil and weather are suitable for this heat-seeking vegetable. Most corn varieties need 120 frost-free days to reach maturity.

"We vegetable growers, we sit on expensive real estate," his father observes. "The government has to take a chance on helping the young farmers who can't afford to buy land. Whether you like it or not, a farmer is somebody."

Jensen Farms Taber
Allen, David and Susan Jensen

It began as three acres of corn that Glen Jensen planted to feed his family. "Dad took some corn to the Crowsnest Pass to sell in 1976," sixtysomething David Jensen reminisces. "That year, a crew was building a highway next to the farm. They started taking our corn on their plane flights home. When we saw there was a market, we said, 'hey, maybe we should grow and sell this stuff!' The patch grew, from three to four, then five acres, to where we are now."

When David and his brother Jim joined the family business, so did their wives. The women made daily trips to Calgary in two trucks loaded with corn sacks, logging sixteen-hour days during the eight-week corn season in those early years. "Direct sales just happened," David says. "Once we got into it, we developed a marketing plan. Now we staff every stand from Calgary south personally, about twenty of them, and have stands all over the province, maybe another thirty or forty."

Jensen Farms is in its third generation. The farm is a big project; with fifty staff at peak, it requires a lot of management. David runs the wholesale arm, his wife, Susan, manages the retail, his nephews operate the bagging line, and son Allen supervises the field operation of two hundred and sixty to three hundred and thirty acres of corn each year. They wholesale to large and small stores, and retail at farmers' markets and all those roadside stands. "Labour's the biggest challenge, there's no one with experience, but we do have some retailers whose families have sold our corn for ten years," he says.

According to David, president of the Alberta Corn Growers Association, Taber's landscape and climate are the closest Alberta comes to California's—if the frost keeps itself in check and waits for those four magical months of sun. "All we need to do is add lots of water," he says. He calculates that one thousand acres in Taber's rural municipality are sown to corn. Each acre produces two hundred bags of corn; each sack is stuffed with forty-eight cobs. Each summer, nearly ten million ears of edible sunlight travel across western Canada. Those familiar burlap sacks appear in farmers' markets, roadside stands and the back ends of pickup trucks from Cranbrook, BC, to Brandon, Manitoba.

Like any food, corn is directly influenced by its terroir—soil, aspect, minerals, sun, light, temperature, weather. Corn likes lots of heat, sandy loam, plenty of water, and one hundred and twenty frost-free days. Few regions in Canada can meet those specific needs. In the sun-baked land surrounding Taber, those tangibles come together to produce corn that is sweet, tender, crunchy, and highly sought after. Taber corn is famous, and even lied about. Every summer, someone at a vegetable stand beside a hot and dusty highway fibs and claims their un-pedigreed corn is from Taber.

"It's our biggest problem. We call it 'counterfeit corn,' and frankly, there are a lot of bad roadside stands," David says. "The safest is one that the producer owns himself." He says that roadside shopping requires nothing more arduous than consumer awareness. Stands must be able to provide provenance: Taber corn growers all produce a certificate of authenticity that includes the farm's name and phone number.

"The corn growers' association owns the 'Taber Corn' label, and people who are members can use that logo. On our Jensen Farms certificate, we use a stamp as a corporate seal; like Braille, it is raised and cannot be reproduced. The buyer can feel it." He suggests that shoppers deal directly with the farmer where possible. "It is easy for wholesalers to slide into selling BC or Washington corn and claim it's from here, especially if we have another cold season, or hail like we've had two years running."

David, at five-six, is not tall, but his corn is. In 2008, his Vision corn won the annual Taber Corn Fest corn tasting. "Every day it's on, we eat corn," he says cheerfully. He's not alone.

All ears are not created equal

Corn varieties have changed drastically since David Jensen's childhood. "Cooks used to put the pot of water on to boil before they went to the garden," he says. That was because older corn varieties converted their sugars into starch minutes after picking. Older hybrids include two-tone Peaches and Cream, Golden Bantam and white Silver Queen. But modern North American open pollinated sweet corn has been cross-bred to produce "supersweet" corn.

"Supersweets'" have high sugar content and crisp skins. Supersweet and lushly textured "sugar-enhanced" varieties are copious, and many have beguiling names that suggest the illicit pleasure of a stripper or porn star. Kandy Korn or Butter Treat, anyone? Bi-colour varieties like Double Delight or Honey 'N Pearl are usually clearly named. Don't ask what an inventive filmmaker dealing in skin would make of names like that. The newest generation, "extra tender supersweet" corn, is often still unnamed, indentified only by a number, and can be two-tone or solid gold. "They have skins so tender," David says, "that toothless old-timers can eat them." Those tender kernels require gentler picking machinery, but he says the ears have longer lives, on the stalk and in the consumer's hands.

vegetable pakoras with mint chutney, cucumber raita and lemon ketchup

These irresistible little fritters are too rich to be dinner all by themselves, but go ahead and try! Serves a crowd.

1 ½ cup (375 mL) chickpea flour
1–1 ½ cups (250–375 mL) buttermilk, plus more as needed
1 jalapeno, seeded and minced
1 Tbsp (15 mL) minced fresh mint or cilantro
kosher salt to taste
½ tsp (2.5 mL) curry powder or ground cumin
1 Tbsp (15 mL) olive oil or cold-pressed organic canola oil
½ tsp (2.5 mL) lemon juice
1 large baker's potato, boiled and then thinly julienned
½ onion, thinly sliced
½ medium zucchini, coarsely grated
¼ cauliflower, small florets, blanched
1 Tbsp (15 mL) minced garlic
sunflower or safflower oil, for shallow pan-frying
Mint or Cilantro Chutney for dipping (recipe follows)
Cucumber Raita for dipping (recipe follows)
Lemon Ketchup for dipping (recipe follows)

Measure flour, buttermilk, mint, salt, spices, oil, lemon juice and garlic into a large bowl. Mix gently with a fork, just enough to blend. Add vegetables. Mix only briefly, until just blended.

Heat oil to a depth of half an inch (1 centimetre) in a wide and shallow pan. Monitor the temperature with a candy thermometer. The optimal temperature is 360°F to 375°F (185°C to 190°C). Alternatively, drop a cube of bread into the oil. When it browns promptly, the oil is hot enough to minimize absorption. Set a paper-towel-lined plate or tray close by.

Using a pair of forks, drop a flat spoonful of batter carefully into the hot fat, as close to the surface as possible to avoid splashing. Turn each fritter with tongs as it becomes golden, then remove and let drain on the plate or tray. Do not overcrowd the pan; a single layer, with room between the fritters, is ideal to maintain an even temperature. Serve hot with dips.

fresh mint or cilantro chutney

This is refreshing made with either mint or cilantro. Serve with any hot dish, curry, or seafood and lamb dishes. Makes 1 cup (250 millilitres).

1 cup (250 mL) packed cilantro or mint leaves, stalks discarded
1 green onion
1 bunch flat-leaf parsley
2 jalapeno peppers, seeded and coarsely chopped
2 garlic cloves
2 fresh ginger root slices
½ cup (125 mL) orange juice
¼ cup (60 mL) lemon juice
1 tsp (5 mL) ground cumin or garam masala
1 Tbsp (15 mL) sugar
kosher salt to taste
½–1 cup (125–250 mL) water

Purée all the ingredients in a food processor.

cucumber raita

Use yogourt that does not contain gelatin. Otherwise the whey won't separate and drain. Draining the yogourt and squeezing the grated cucumber creates a rich and creamy dip that won't "weep." Serve this cooling salad with spicy dishes or use it as a dip. Makes about 2 cups (500 millilitres).

3 cups (750 mL) unflavoured yogourt
1 Long English cucumber, coarsely grated
2–4 garlic cloves, minced
kosher salt and freshly ground black pepper to taste
2 Tbsp (30 mL) minced parsley or cilantro
2–4 Tbsp (30–60 mL) toasted coconut (optional)

Line a fine mesh strainer with a damp kitchen towel and dump yogourt into the sieve. Let it drain for 30 to 45 minutes. Discard the whey. Use your hands to squeeze the grated cucumber dry. Mix yogourt in a bowl with remaining ingredients and season to taste.

lemon ketchup

Use as a dip or garnish for sautéed vegetables, fish and meats. Makes 1 cup (250 millilitres).

2 lb (1 kg) ripe tomatoes, coarsely chopped
zest of 1 lemon
3 Tbsp (45 mL) extra virgin olive oil or cold-pressed organic canola oil
kosher salt to taste
½ cup (125 mL) lemon juice
2 Tbsp (30 mL) pomegranate molasses
freshly cracked black pepper to taste

Combine all ingredients in a heavy-bottomed pot and simmer uncovered for 30 to 40 minutes, until soft and thickened to dip consistency. Taste and adjust seasoning as needed.

The ancient skill of seed-saving, growing onions and garlic, among other vegetables, for a second season to allow them to produce seeds for subsequent seasons, often results in eerie-looking produce. This practice is increasingly important in the face of seed patenting.

Mark Gibeau is an artist and metal sculptor as well as a grain grower.
His choice of metal motif for his farm is an elaborately simple wheat sheaf.

w is for wheat

Heritage Harvest Strathmore
Henry Winnicki, Mark Gibeau, and Ray LeFebvre

Elaborate metal wheat stalks tower above Mark Gibeau's farm's gate in a perfect cut-out against the sky, a deliberate choice of motif for an artist and heritage wheat grower. Mark lives on one of three adjoining farms near Strathmore that are home to Red Fife wheat and its descendant, Marquis wheat, and a handful of other heritage grains. "Old wheat and old-timers have intrinsic value," Mark says. "Old farmers have an immense amount of knowledge. Sure, it takes finesse to get them to talk and remember. Many were organic without the name for far longer than they have been conventional farmers, going back before they started using chemicals after the war."

Mark, Henry Winnicki and Ray LeFebvre all wanted to go back to farm life. Mark ran the glass-blowing facility at Alberta College of Art and Design (ACAD) in Calgary; Henry and Ray were part of Calgary's oil and gas business. "But financially, we couldn't run with the big boys," Mark shrugs. "We needed to find a niche crop in short supply." The trio's research led almost inescapably to organics, and from there to old grains. Mark learned that Red Fife doesn't grow or mature like the current wheat strains. He redesigned his air seeder and adapted a harrow into a tine weeder with long tines for what he calls "post-immersion weed control" after the crop is sprouted, and they rely on biodynamic compost tea as the backbone of their soil fertility program.

"Red Fife will never make Wonder Bread," Mark says as he shows me a bucketful of reddish wheatberries. "It has a different taste, texture, feel and smell than modern wheat, and what I call 'my Gran's house baking' aroma. And there is that significant movement back to traditional foods."

Historian and wheat curator Sharon Rempel's entry about Red Fife in the Canadian Encyclopedia says that Canada's oldest wheat had its origins in Peterborough, Ontario, when farmer David Fife planted red seeds sent to him by a friend in Glasgow. The friend had collected the seeds from a Ukrainian shipload of wheat. By the 1860s, settlers across Canada were planting Red Fife given to them as incentive gifts by the CPR and the federal government. In 1876, the first shipment of Red Fife went up the Red River on the *Minnesota* to eastern Canada's markets. Eight years later, one thousand bushels went to Britain. Red Fife was the dominant wheat until 1900, when it was replaced by its descendant, Marquis wheat.

Until the late 1980s, Red Fife was grown only by plant breeders for their seed collections, but Slow Food "Ark" activists have taken an interest in restoring Red Fife. Mark and his colleagues are part of the wave. In 2010, they harvested one hundred acres of Red Fife. In 2011, they set up Alberta's first farmstead stone-ground mill on Mark's farm. The mill's circular stones, quarried in France, are two feet in diameter, and they slowly move around a three-foot "well" to grind wheat into flour without generating any heat that would degrade the flour. The process is remarkably similar to olive presses that produce cold-pressed olive oil.

Mark describes the mill's geographic location forty minutes east of Calgary's one million residents and potential customers as "dumb luck," adding pragmatically, "Not being able to keep up with demand would be a good problem."

Ehnes Organic Seed Cleaning Ltd. and Back 40 Organics Ltd. Etzikom
Bernie Ehnes

The first story Bernie Ehnes tells me is about driving the grain truck taking wheat on the run from the big Massey combine as it rolled down the field. His older brothers Eddie and Richard were bringing in the harvest on the family farm, and they needed another pair of hands. It was 1965, their father was ill and even a seven-year-old could help.

Most of the other stories include Brian, his twin brother. "In 1994, Brian and I went out and bought a half section jointly. Everything we did, we did together," he says. "We bought tractors, combines, even our first truck when we were sixteen. If you add it up and split it down the middle, it would be half and half. We were comfortable with that." That comfort level continued unbroken until Brian was killed in a traffic accident in 2010. Brian's son Tyler is filling his dad's shoes.

The brothers each received a quarter section of land from their father, Pius, in 1989. The twins broke the sod, planted mustard and considered what to do next. "Brian went to a farm show in Lethbridge," Bernie recalls. "He talked to a mustard contractor who mentioned organics. I hadn't heard much about organics, but Brian came back saying, 'if we do it this way we'll get better money.' Three years later, we had that land certified."

Over several seasons, the brothers figured out new techniques and moved into a fifty-fifty farming operation, with half their land in summer fallow, and half seeded in durum, spring wheat, barley, three varieties of mustard, plus flax or buckwheat. In 1996, they started a custom cleaning plant on the home quarter, where Bernie and his family lived.

Late in the year, a grain auger's power takeoff took Bernie's left arm above the elbow, broke four ribs and his scapula, and tore ligaments on his left knee. "We hired men to fill my gap at the seed-cleaning plant and on the farm," Bernie says quietly. "I was in rehab. It took a while." He wears a mechanical prosthesis when he needs to, and carries welding magnets in his toolbox. "If I'm hammering and using a chisel or punch, I use welding magnets to hold my chisel or punch. When I drive grain trucks I use my prosthesis, but hardly ever for the combine or tractor."

They narrowed the crops to wheat and barley, and the brothers realized they needed a bigger land base in 1999. Ten years later, historian Sharon Rempel called Bernie, looking for a grower to preserve the historic Red Fife wheat that Canadian settlers received between the 1860s until 1900. "She was concerned about the seed being kept as pure as possible without blend-ing or cross-breeding," he says. They planted ten acres with the seed Sharon sent, and took off two hundred and sixty bushels. The next crop filled one hundred and twenty acres, and yielded over three thousand bushels. As of 2011, that wheat was in a grain bin on the farm, waiting to be sold to other growers, mills, and the occasional home bakers interested in heritage wheat. "I can wait," Bernie says. "It can keep a long time; grain doesn't go off."

The bulk of Bernie's grain goes to BC, for use in organic breads and flours. Some of the barley Brian and Tyler planted in 2010 is malted for the burgeoning microbrewery trade. "It makes me feel good that we are

Mark Gibeau observes that Red Fife wheat's nutty flavour makes bread that will never be mistaken for factory-style mass-produced bread.

producing good-quality products that are feeding people," Bernie says. "Someone somewhere is eating healthy because that's the way I grew up. We need to see the relationship between eating and health, and get back to a slower-paced lifestyle. It's lower stress."

Since his twin's death, Bernie, now in his early fifties, has had time to adjust. Despite the irreplaceable loss, the farm will carry on, he says, with Tyler working beside Bernie. "We'll simplify some of our marketing. Tyler is in his mid-twenties, he's getting into it, Brian was starting to show him so many things. He's a good head. When Tyler drove his first combine, he was as young as I was."

flaxseed and oat bread roll with basil, gouda and cold-pressed canola oil

Chewy texture in a slightly rustic style make for a wonderful light lunch or brunch accompaniment and a hearty aside to any soup or stew. Makes 4 loaves.

dough:
1 Tbsp (15 mL) yeast
2 Tbsp (30 mL) white sugar
¼ cup (60 mL) warm water or milk
2½ cups (625 mL) all-purpose flour
1¼ cups (310 mL) whole wheat flour
1½ cups (375 mL) rolled oats
kosher salt to taste
2 Tbsp (30 mL) flaxseed
2 cups (500 mL) hot water or milk
1–2 Tbsp (15–30 mL) cold-pressed canola or flaxseed oil

fillings:
1 cups (250 mL) diced and seeded tomatoes
4 Tbsp (60 mL) chopped fresh basil
4 garlic cloves, sliced
2 Tbsp (30 mL) chopped fresh parsley
1 tsp (5 mL) minced fresh rosemary
2 Tbsp (30 mL) minced fresh thyme
3 green onions, minced
1 cup (250 mL) grated old Gouda or cheddar
4 Tbsp (60 mL) cold-pressed canola oil
kosher salt and freshly cracked black pepper to taste

In a countertop mixer, combine yeast, sugar and warm water or milk. Let this mixture stand for about 5 minutes, until it is puffy. Add flours, oats, salt, flaxseed and hot water or milk. Mix with the dough hook until it is a smooth ball, about 5 minutes, adding flour or water as needed. Turn the dough onto the counter and knead by hand until smooth and soft. Oil the bowl's interior with the canola or flaxseed oil, then roll the dough in the bowl to coat it lightly with oil, and put in a warm place to rise.

Preheat the oven to 425°F (210°C). When the dough has doubled in bulk, punch it down and divide it in four equal pieces without kneading it. Shape into four ovals about 12 x 6 inches (30 x 15 centimetres), using the palm of your hand to shape and flatten the surface. Place on parchment-lined baking sheets and evenly divide the diced tomatoes, basil, garlic, herbs, green onion and cheese among the four pieces. Drizzle each round with oil, then sprinkle with salt and pepper to taste. Roll up snugly. Bake the loaves until nicely brown and crusty, about 20 minutes. Serve warm or at room temperature.

Cook's note: Alternatively, either flatten the dough into 4 rounds, add toppings and bake flat, or mix the toppings into the dough and shape into boules, or round loaves.

x is for xeriscaping experts, nearly extinct (plains bison)

Olson's High Country Bison, Spread Eagle Ranch and High Country Ranch Waterton and Bragg Creek
Tom and Carolyn Olson

A swathe of stars glitters across the early-morning canopy of winter sky as Tom Olson drives from Bragg Creek to Calgary. Tom is a tax lawyer by day and a bison rancher by night. What looks like paradox is in fact connected. Tom is a weekend warrior committed to reclaiming the native range by the simple expedient of allowing bison to live as they always have. "I love spending time with Jewish tax lawyers back east," Tom says. "Everything, including meetings, happens at the dinner table. When I asked, one guy told me, 'It's our culture: we have big meals, we debate the Talmud.' I love it, restoring food and the family meal to their rightful intellectual and religious experience."

Tom, a Mormon now in his fifties, studied biology, then law, intending to become an environmental lawyer. "But in the '70s, in Canada, no one knew what an environmental lawyer did. So we've done our environmental cause on the side."

The bison, four thousand of them on forty-seven thousand acres spread over five ranches, are a family project. "We couldn't have succeeded without the kids, their full buy-in and all their help," Tom admits. "Help" means building fences, managing crews, running the irrigation project at the High Country Ranch near Bragg Creek, and working with the bison. The ten kids, preteens to thirtysomethings, all home-schooled by Tom's wife, Carolyn, were raised far from "city kids with nothing to do." Tom jokes, "We taught our kids to be independent on the ranch, and my wife blames me, now that our oldest is an engineer in Beijing."

Tom grew up in the Porcupine Hills, south of Calgary. In 1985, the couple bought the Bragg Creek ranch. For health reasons, Carolyn

Paradoxically, the best way to save bison from extinction is to eat them, says Tom Olson. To do that, they walk with the bison, become a part of their matriarchy, keep them wild and handle them minimally.

ruled out raising beef, then they briefly considered muskox and yaks, but discarded both as not native to North America, buying six bison instead. When he and Carolyn started to acquire ranchland, they deliberately chose places with ecosystems friendly to bison—foothills parkland; montane near Waterton; foothills fescue; dry mixed-grass prairie in the Cypress Hills; and boreal forest and aspen at Pine River, Manitoba.

They populated each ranch with the animals. "We wanted the bison to fulfill their traditional role. They were the keystone species, on whom all the other species were dependent. Bison hooves and grazing patterns formed the landscape. Now we have the prospect of bringing back the insects, birds, even the predators, all as integral parts of the system we envision restoring."

The goal was raising healthy food on a healthy landscape where nature does most of the repair work. "If we could change the world, I'd have all those people who raise industrial food in feedlots and factories spend a day sitting with the bison." Toward that aim, Tom invites chefs from around Alberta to visit, observe the animals and eat the less trumpeted cuts; as with any endangered species, to save it, you have to eat it.

It's a big dream. When Tom attended Slow Food's Terra Madre in Torino, Italy, in 2010, he observed the interconnected objectives of delegates. "Slow Food is an umbrella for many noble causes, all underpinned by food and its vital role in society. On this continent, we need that food revolution most," he says. "It will be led by a bunch of like-minded Slow people who have thought carefully and feel strongly about supporting that view—not by the government, nor large powerful organizations. It's percolating, a great debate is happening amongst intellectuals, and some of them are chefs."

The stars fade as Tom approaches Calgary. "Bison are the ethical eco-friendly alternative to beef. The conflict between cattle and bison ranchers comes from two bases. One, that some ranchers are rednecks, intolerant bigots who want everything their way, who think they can destroy property and walk over rights. And two, even people who aren't redneck bigots feel threatened by things that are new." Consumers have choices to make. "If you vote with your dollars, you get socially conscious and quality food."

Four Creeks Ranch Silver Valley
Ted Buchan

In high summer I walk the high escarpments on the south slopes of the Peace River in West Dunvegan Provincial Park with bison rancher and environmentalist Ted Buchan, who totes a rifle in case of bears. We see evidence, but no bear. After squatting to examine fescue and prickly pear cactus, I spot shadows in an aspen grove that turn out to be a cow moose and her calf.

En route to the river bluff, Ted bounces us through Four Creeks Ranch in Silver Valley, across the grazing lands of his bison. We navigate an ingeniously simple system of gates and chutes, all controlled by ropes strung from a central tower so that moving bison can be easily and calmly directed by one pair of hands. "Bison dislike small spaces and crowding," Ted explains, so he moves them over several days in progressively smaller groups into successively smaller paddocks to keep stress levels low. By then, we are standing across the fence from a bull weighing nearly eighteen hundred pounds. We watch it drop and roll in the dust. Rutting season is near, and a few minutes later, the same bull charges and chases a smaller bull from his herd. The younger animal leaps a six-foot fence. Ted nods at my astounded look. "Yep," he says. "Six feet is easy for them."

A rousing raconteur who earned his stripes on oil rigs in his youth, Ted tells tales of his past as we roll slowly through the herd, pausing to watch yearlings avoid the bull. Four Creeks Ranch abuts Peace River Wildland Provincial Park, west of the town of Peace River. The ranch was certified organic in 1991, but Buchan has been "green" since age thirteen when he witnessed his childhood waterway, the St. Lawrence Seaway, becoming a toxic soup. He attended Slow Food's inaugural Terra Madre, a global

Many ranchers, including Ted Buchan, are environmentalists and naturalists with a deep appreciation for the natural order.

assembly of artisanal producers in 2004 in Torino, Italy, and spoke there about his low-key approach to ranching bison. "I got mobbed every time I showed up in my cowboy hat," he drawls. "It was the North American tribal dress."

The provincial park adjacent his land and the Peace River is in place partly due to Ted's efforts. It protects a unique ecosystem—including the native shortgrass called fescue, prickly pear cactus, and the river itself—from hungry oil and gas companies. Eighty per cent of the families in his region are employed by oil and gas, he says, and nearly all showed up to join the local branch of the environmental protection group when it formed.

Ted returned to Torino in 2008. "I shed a bunch of preconceptions about food on my first trip to Italy, my first trip off the continent," he says. "We all belong to organizations, we all attend meetings. Terra Madre and the Slow Food movement spell something larger. I'm a hermit, it's hard to be around so many people at a time as I was there, but it recharges my batteries. And as part of a small group in Alberta, I feel like a voice in the wilderness. It helped to see that the Slow movement ties in perfectly with 'organic' and 'environmental.'"

The biggest change to come from Terra Madre so far is an increase in awareness of food security issues at a grassroots level. "It's all tied together; petroleum and the food system are so linked, and the economy is part of that. High oil prices precipitated the crash. I wonder how people will get affordable food after cheap oil ends."

Buffalo Horn Ranch Olds
Peter and Judy Haase

> "The greatness of a nation and its moral progress can be judged by the way its animals are treated."
> —Mahatma Gandhi

Bison is one of my favourite meats. A mouthful of its tender, juicy richness is inextricably bound to a bus tootle on a cool day to a pasture north of Calgary, where I observed a herd of bison peacefully grazing. The clouds congregating, my friends were laughing as we knelt in the grass to get a closer view of the herd. Each time I braise bison short ribs, that summer day lights my kitchen, regardless of the season. "Every fridge should have stuff like this in it," my son says one evening as he eats another plate of ribs and mashed potatoes. He's right.

One of my favourite sources of grass-raised bison is Buffalo Horn Ranch, the site of my epiphany. Co-owners Peter and Judy Haase left their lives in Calgary, as a professional photographer and a teacher, in 1994 to raise bison. Peter's family had a nine-hundred-year history as German farmers and Judy's people were Ukrainian farmers. "We both have the recessive food-producing gene," Peter says sardonically. It's not a gene that pays well, he adds. "The bank laughed us out of the building when we said we wanted to start a bison herd." So they sold their house. By 2000, they had acquired a ranch near Olds and a herd of one hundred cows.

They didn't count on the 2001 market collapse, drought and BSE in 2003, or the drought and recession of 2008–09. The herd now numbers forty-five cows and their calves, who live in the matriarchal herd for about

Peter and Judy Haase enjoy the outdoor aspect of their ranch life, and have strong opinions about the necessity for humane animal treatment.

three years—all told, one hundred and fifty animals—what the pair can comfortably manage, and enough to ensure a living.

The Haases are content to stay small. "I'm no longer concerned about closed borders like we had with BSE," Peter tells me. "We have a dedicated customer base, and with a growing shortage of bison, the demand will only strengthen. Land in Alberta is too expensive for us to expand, and we don't want to work so hard in production and marketing."

Their tall, lanky frames are familiar sights at several farmers' markets in the Calgary area, and at restaurant doors. "Face time with people," Peter says. "We've tried advertising and it hasn't done much for us." What does work is selling directly and conversing about how their animals are raised and how to cook lean bison.

Slow Food's philosophy of connecting growers with consumers has shaped their approach. They attended the organization's biennial gathering, Terra Madre, in 2006 and again in 2008. Their first trip was exciting and relieved the sense of isolation many sustainable growers have in Alberta, but two years later, Peter and Judy felt embarrassment and frustration "at how damaging we can be as people, to each other and to the land. There's less hope, and the same issues."

They continue to believe in small things as agents for change. Cook at home. Grow things. Share. Sit down and eat with others. Visit a farm. "Two systems are needed, one for export, one to feed Albertans," Peter says. He talks hopefully about a system that would educate new farmers, and provide access to land for those who want to grow food. "We should take the economics out of the equation when we talk about feeding Albertans. It needs to be about food security—good, clean and fair food for Albertans, and the world market will dictate what we do with our surplus." Their biggest concern is climate change. "Over the past decade we have seen so much goofy weather, freak storms, droughts. The carrying capacity of our land is much lower than it was in the past due to lower rainfall. If it doesn't rain, you can't make money in agriculture."

Valta Bison Farms Valhalla Centre
Darlene and Gil Hegel

Eight hundred kilometres is a long way to go for a piece of pie. But the trip is worth the gas tax and the time. Northwest of Edmonton in the heart of Peace Country, I eat berry pie with Gil and Darlene Hegel, at Melsness Mercantile in Valhalla Centre. I could have had the bison and conversation a lot closer to home, as the Hegels' business, Valta Bison, has several retail shops in Calgary. But then I wouldn't have enjoyed the blueberry cream pie, nor the sighting of the Peace River's newest provincial park, the drive through a herd of ruminating bison, or the trip on the Shaftesbury ferry across the Peace River at sun-up.

Over pie, Gil and Darlene Hegel tell me that half of Alberta's bison is raised in the Peace Country. What matters is re-establishing a community to support and nurture farmers, who support and nurture the community in turn. "People want to be recognized," says Gil, who has learned to enjoy the chat and interaction that is part and parcel of being in retail. Darlene, no mean cook, says it also gives them a chance to educate people about just how to cook their meat.

"I like the animals," Gil says quietly. He has lived the life of a full-time farmer. Gil and Darlene were raised in the Peace Country and are proud

Young bison live within a matriarchy, where the boss cow rules. Like most animals, they possess an uncanny ability to sense frustration in humans, and require calm, unhurried handling, usually on foot.

of their region, and their family history. Darlene loves to recount her Norwegian family's settling of the home farm, and the family's connection to the Valhalla Centre Mercantile. Their soft voices and demeanours are protective colouration, concealing ancestral bedrock and business acumen.

The pair set up shop in 2004 in the Calgary Farmers' Market and stayed several years, attracted by Calgary's boom and the market's central location. In 2007, they translated a growing storage need into a new retail store in Ramsay, in inner-city Calgary. This old district has a sense of history. Gil laughs and says, "It has a 1911 front with a 1945 house on back, and I'm sure the building was up to code in 1945." Natural and smoked sausages are the core of their business, including Mennonite, smokies, sun-dried tomato, garlic coil, and pepperoni sticks. At the shop, a small child comes by and buys his first-ever "I bought it alone" bison pepperoni stick. Darlene gives him two to mark the occasion. These Peace Country ranchers have made a warm name for themselves, based on their successful and low-key marketing of bison as an un-intimidating meat, one stick of pepperoni or slice of pie at a time.

braised bison hump with cherries and juniper

Braised meats dry out easily, so keep the bison covered during cooking, and afterwards, as it cools. Substitute lamb shoulder, pork butt, or beef chuck or short ribs if you wish. Serve with roasted root vegetables and potatoes. Serves 4–6.

3 lb (1.5 kg) bison hump or chuck, tied if necessary, patted dry
kosher salt and freshly cracked black pepper to taste
2–3 Tbsp (30–45 mL) extra virgin olive oil
2 yellow onions, chopped
1 carrot, chopped
6 garlic cloves, chopped
2 Tbsp (30 mL) red wine vinegar
1 cup (250 mL) red wine
1 bay leaf
1 tsp (5 mL) whole peppercorns
10 juniper berries
1 fresh rosemary sprig, minced
1 fresh thyme sprig, minced
1 whole star anise
1 tart apple, unpeeled, cut into ½ in (1 cm) cubes
½ cup (125 mL) fresh, dried or frozen sweet cherries
4–6 cups (1–1.5 L) roasted chicken, veal or beef stock
4 Tbsp (60 mL) minced fresh parsley

Preheat the oven to 300°F (150°C). Season bison with salt and pepper. Heat oil in a heavy-bottomed frying pan and sear meat on all sides. Add onion, carrot and garlic, letting vegetables brown too. Season vegetables with salt and pepper. Place meat and vegetables in a large roasting pan and add all remaining ingredients except parsley. Bring to an active simmer on the stove. Cover the pot's contents with a piece of parchment paper, fitting it snugly onto the surface of the meat and liquid. Cover with a snug lid. Cook in the oven for about 3 to 4 hours or until meat is fork tender, turning it with a pair of tongs several times during the cooking process.

Remove the bison from the liquid and wrap it in foil or plastic. Put the liquid in its pan back on the stovetop over medium-high heat and boil until it is reduced to sauce consistency, skimming to remove any fat. Strain if you wish. Thickly slice cooked bison meat and add it to reduced stock along with parsley. Heat thoroughly and season if needed. Serve.

y is for yogourt

Vital Green Farms Picture Butte
Joe and Caroline Mans

The narrowest of an opening on the inside rail was all that Joe Mans needed to make a name for himself as a dairyman in Alberta. Without that gap, Albertans would not be sighing over Vital Green Farms' luscious yogourt and 52 per cent heavy cream.

The gap, Joe says, was Alberta Milk's policy of setting aside a percentage of quota for farmers interested in producing organic milk but who were without the funds to pay the steep price to purchase quota. (See Milk quota, p. 119, and About quota, or chicken by the numbers, p. 27.) The policy gave Joe and his wife, Caroline, the chance to lease a quarter section of farmland near Picture Butte—but only if 75 per cent of the milk they produced was organic.

"It's hard to get into the dairy business," he says. "We wouldn't have otherwise. Any young kid can't, plain and simple. Banks like 50 per cent down on a loan to purchase quota." These days, that's close to forty thousand dollars per cow, so a herd of fifty cows is worth two million bucks in quota. "How can you get a million-buck down payment together? Cash was the biggest challenge then, and it still is." For all its drawbacks, Joe is quick to admit the quota system also guarantees a regular cheque and offers control over trade tariffs. "Without quota, Canada would become like our big southern neighbour, and no one would make any money."

Joe moved to Canada with his Dutch farm parents and most of his siblings in 1982. Dutch farmland was shrinking, absorbed into urban centres and nature reserves. "You can't eat a nature reserve," Joe says tartly. They arrived with little cash and a lot of hope, and Joe promptly married a Dutch woman he had met in Canada in 1980 on a student work experience trip. Three years later, he and Caroline started the province's second certified

Herbed yogourt is a delicious accompaniment to vegetables as well as fruit, granola and grilled meats.

organic dairy with a herd of fifteen Holsteins, chosen specifically for their butterfat content and yield.

"We decided to do on-farm processing too—no one else was, and it let us add value before the milk left our farm," Joe recalls. He rented the milk plant vacated by Ben and Anita Oodshorn when they relocated their nearby goat dairy to a larger farm. The plant, four miles away, was in the heart of a Dutch enclave that became increasingly important as Caroline began to bear and raise their children.

Joe pasteurizes his Holsteins' milk and separates the cream, makes yogourt, crème fraiche, cream cheese and chocolate milk. "Nothing gets homogenized," he says, "it is not a good thing." Homogenization, which is the norm in the dairy world, is the process by which miniscule cream particles are suspended in lower-fat milk. Some people believe that straining the fat through tiny pores under great pressure increases risk of rancidity and oxidation.

The farm grew to one hundred and fifty acres, and the herd to about fifty cows. Caroline had her eleventh child in 2006. By 2009, when their own milk plant was completed, Joe himself rarely milked anymore. By then, the older boys and part-time hired help were milking the cows.

"The lack of a work ethic—it is not just a Canadian problem," Joe says. "In Europe too, cities are too big. City kids come home from school, they sit in front of computers. On a farm, kids help pick rocks, clean the yard, move straw. There is always something to do, and kids learn how to work." With eleven kids in the family, life is full of small but important details. "We go to church. School is very important. We all drink milk."

He means raw milk. As the dairy's owners, the Mans family are the only ones who are legally allowed to drink raw milk. The fines are substantial for violations, Joe says, and the risks include loss of the dairy licence. Like quota, the topic of raw milk has Joe nodding as he acknowledges the pros and the cons. "It's healthy but not allowed to be sold. You understand me." From his spot on the inside rail, Joe Mans isn't about to risk what he has worked so hard to achieve.

Bles-Wold Dairy and Bles-Wold Yogurt Lacombe
Tinie Eilers and Hennie Bos

It began as an energetic mother's proactive decision to make a healthy, sugarless yogourt for her diabetic daughter's breakfast. It burgeoned into a thriving business. Along the way, Tinie Eilers became Alberta's pre-eminent yogourt maker. "You do what you need to do and make things better for the kid. We have had lots of luck in our life, and we believe that work is healthy. But if you do what you like, work doesn't feel like work," Tinie observes.

Both from dairying stock, Tinie and Hennie, and their two children, emigrated from the Netherlands in 1994, when Martine was thirteen and Gerard was nine. "Sure," Tinie says, "The Netherlands is heavily regulated, but imagine sixteen million people living between Calgary and Edmonton. In those circumstances, it is fair to have all those regulations. There are fewer people here, so it's easier. There are more opportunities."

Hennie bought sixty Holstein cows, and in 1996, Tinie, a knitter, seamstress and gardener with the skilful hands of a born craftswoman, wrote to her brother in the Netherlands, asking for his yogourt recipe and a pasteurizer. Martine liked the result of her mother's kitchen experiment. So did Hennie and the neighbours. Tinie took her extra yogourt to the farmers' markets in Ponoka and Lacombe. "The manager from the local Co-op store came to us, and Bles-Wold Yogurt made the leap to grocery stores."

"It was exciting, we got lots of help from Alberta Agriculture; we didn't know where to look for information in the beginning! They made it all easy—the regulations, labelling, advertising, everything. Some days I worked fifteen to sixteen hours, so it wasn't all that easy, but I liked it. It didn't feel that I was working, just go, go, go."

The yogourt is processed on the farm, in a federally inspected building, within hours of milking time. Milking and the dairy's management are now in the hands of Martine's husband, Ben Varekamp, and they will one day take over the farm. "We are currently in partnership, but that will change. It opens doors for us," Tinie says. Her husband is deeply involved in dairy farmers' organizations at the federal and provincial levels, and travels quite a bit. The herd is up to two hundred and seventy milking cows, crossbred Holsteins and Fleckvieh cows, a dual-purpose breed (for meat and for milk) from Bavaria. They produce forty-five hundred litres (about one thousand gallons) of milk each week, which makes an equivalent volume of yogourt.

Since 2005, the cows have been fed ground sunflower seeds to increase the milk's naturally occurring conjugated lineolic acid, or CLA. This essential fatty acid is shown to provide health benefits related to cancer, heart disease, kidney disease, diabetes, bone density and obesity. Tinie is particularly aware of wellness, and not simply because of her daughter's diabetes. Hennie has survived a pair of cerebral strokes, and Tinie laughingly says he has more energy than she does. "The best thing is being healthy and happy," Tinie says. "It means you can do what you want to do."

yogourt tiramisu alla dennice with fruit wine or mead zabaglione

I have re-imagined this classic Italian dessert, abandoning the traditional ladyfingers in favour of diced brownies and Italian cookies—*amaretti* or *cantucci*—and I have replaced part of the cream cheese with drained yogourt "cheese." It's still a celebratory dessert. The zabaglione sauce is delightful made with fruit wine, *vin santo*, late-harvest or ice wine in place of the traditional Marsala, muscat or champagne. Use the sauce as a foil for seasonal fruit, biscotti or tuiles to celebrate special occasions. Serves 8.

cheese & cream:
3 cups (750 mL) gelatin-free plain yogourt
2 cups (500 mL) cream cheese
1 cup (250 mL) ricotta cheese
4–6 Tbsp (60–90 mL) whipping cream
½ lemon, juice only
1 Tbsp (15 mL) honey
zabaglione:
4 large egg yolks
¼ cup (60 mL) white sugar
¾ cup (180 mL) port-style fruit wine, sweet mead, vin santo,
    late-harvest or ice wine
½ tsp (2.5 mL) freshly grated nutmeg
the rest:
2 cups (500 mL) crumbled amaretti or cantucci
4 brownies, cut in ¼ in (½ cm) dice
1 cup (250 mL) espresso or very strong coffee
½ cup (125 mL) Kahlua, or more as needed
1–2 Tbsp (15–30 mL) cocoa, for dusting
1 tsp (5 mL) cinnamon, for dusting

Line a fine mesh sieve with a clean kitchen towel. Pour yogourt into the sieve and let rest over a bowl to catch the whey. Let yogourt drain for 45 minutes, about 1½ cups (375 millilitres) yogourt should remain. Discard the whey. Beat the cream cheese in a large bowl until smooth, using the paddle attachment of a stand mixer or a vigourously employed wooden spoon. Add drained yogourt, ricotta, whipping cream, lemon juice and honey. Stir until smooth. Set aside.

To make the zabaglione, vigourously whisk egg yolks and sugar in a medium-size stainless steel bowl until pale yellow, thick and doubled in volume, about

10 minutes. Place the bowl over simmering water, shallow enough that the water does not touch the bottom of the bowl, and add the wine in a very slow drizzle, whisking continuously. Whisk until the mixture is fluffy and holds soft peaks when dropped from the whisk, about 10 minutes. Remove from heat and set aside.

Evenly layer the bottoms of 8 4-ounce (125-millilitres) ramekins with half the amaretti bits and brownie pieces.

Combine the coffee and Kahlua in a measuring cup. Spoon or pour half the mixture over the brownie-cookie mix in the ramekins.

Spoon half the yogourt-cheese mixture overtop and use a teaspoon to smooth the surfaces.

Add the remaining amaretti bits and brownie pieces, then remaining coffee-Kahlua mixture, dividing each evenly among the ramekins.

Spread remaining yogourt-cheese mixture overtop and smooth with a spoon.

Scoop a tablespoon (15 millilitres) of zabaglione onto each ramekin. Wipe the dishes' edges clean with a damp cloth. Dust sparingly with cocoa and cinnamon. Chill for at least an hour before serving.

z is for zizania (wild rice)

Lakeland Wildrice Ltd. Athabasca
Alice and Wayne Ptolemy

When the quaking aspens weep gold, and the grains of wild rice blacken in their ruddy husks, harvest time has arrived. Wild rice is the original, truly North American grain crop. Not a rice, but a tall grass, its elongated kernel is elegant and eloquent. Even its name sings. *Zizania aquatica* grows where the Precambrian Shield wraps its rocky arms around the grassland, where lakes are encased in the sleeve of glacial rock. Walk to the edge of a lake rich in wild rice, and you may not even see the water's edge until your boot toes nudge the mud.

You have to travel north of Edmonton to see wild rice in Alberta. It's an uncertain crop. Wild rice harvester Wayne Ptolemy says 2010 was the best year ever, but an early frost froze the shallow lakes he leases before he could get out in the airboats to harvest it. "Minus ten two nights in a row in late September, and it was gone," he grumbles. Wayne lives along the Athabasca River, one hundred miles northeast of Edmonton. "You get a good crop every three to four years. But every year I bite the bullet and keep some seed. It's more important to have seed than profit—if you got no seed, you got no crop."

"A good crop" translates into a harvest of one hundred thousand pounds of green rice, with generous losses to the wild ducks and geese. But it begins with water. "It's hard to find water to grow rice," Wayne explains. "You look for healthy lily pads that mean the pH is right. Rice needs shallow water, from two to four feet deep, and it must be crystal clear, no muskeg or alkaline water. Furthermore, in Alberta, you can't lease moving water, but lakes are hard too—most are taken up with non-rice stuff—and if fish spawn there, or if people go to it, the rice won't grow. Plus it can't have a dock, and rice-growing can't disturb the forest."

Airboats have replaced canoes and hand-held winnowing
sticks on Precambrian lakes sown with wild rice.

The wilderness has been Wayne's home and partial source of income for years. Ridges and low lands drain into the Athabasca River, surrounded by spruce, poplar, birch, balsam and jackpine. Wayne has been a boreal forest trapper for twenty-some years, bringing in pelts from lynx, marten, timber wolf, otter, bears, even squirrels. "All except wolverines," he insists.

Wayne always wanted to farm, so when he retired from the oil patch, he and his wife, Alice, moved into the northern bush and started farming wild rice. "I've been broke ever since," he says acerbically. Six feet tall, still muscular, he splits all his wood by hand, five or six cords per year. "I'm in my sixties, [but] I can keep up with any twenty-one-year-old. I like the bush. In the wilderness I can get away and live in Wayne's world, not relying on other people."

There's no rice processing in Alberta, so Wayne ships his green rice to Manitoba. "You can't make a real living at wild rice," he says, sounding apologetic; Alice, his favourite dock foreman, owns a fabric store. "But she threw a lot of bags of rice in her time."

Canadian growers have banded together, labelling their rice as "Canadian Lake Wild Rice" to differentiate it from American wild rice, which is often another variety (*Zizania palustris*). But there are fewer wild rice harvesters than there used to be. Wayne blames the cost of equipment, the unreliable monetary returns and the high cost of insurance, combined with the absence of a processing plant in Alberta. "There's just me and a young Swiss guy left now, after more than a hundred members of the wild rice association in its heyday."

Harvesting wild rice

Wild rice is an annual that reseeds itself. Each fall, the grain falls from its shoulder-high stalk to the shallow lake floor, and lies dormant there until the spring, when it germinates and grows. Some harvesters, including Wayne, re-seed it, so it is not as wild as it once was.

In a canoe, hand harvesters hold wooden winnowing sticks to first fold the stalks across the canoe and then tap the plants so the grains drop into the canoe's bottom. Because the rice does not all ripen at once, it takes multiple trips to collect the rice as it darkens from green to near-black inside its rust-coloured husk. Multiple trips by modern air boats that hover above the water like oversized dragonflies speed the process.

The green rice is hauled to a shelter, where it is spread out to cure and finish ripening, hand-turned daily for a week. The rice is parched in slowly revolving tubs, reminiscent of the business end of a cement truck, over a wood fire to dry and flavour the grain as the husk crisps and crackles. By this point, the rice has lost about half its weight, and the moisture content is one-tenth of what it was at harvest. After the desiccated husk is machine-removed, the grains are graded by their diameter before being packaged. The highest grade is as finely textured as sewing needles.

zizania (wild rice) and cranberry risotto with fennel and sautéed zucchini

Not the norm, no it isn't. The average "normal" risotto uses Italian *arborio* or *carnaroli* rice to create a tender, creamy savoury "rice pudding." This cross-cultural expedition utilizes nutty wild rice, simmered and added to sautéed fennel and zucchini ribbons. Substitute mushrooms for a heartier, woodsy winter version. This is good with grilled fish and roasted meats. Serves 6.

1½ cups (375 mL) wild rice
kosher salt to taste
3–4 Tbsp unsalted butter
1 onion, minced
4 garlic cloves, minced
1 bay leaf
2 tsp (10 mL) minced fresh rosemary
1 red bell pepper, finely diced
1 zucchini, peeled with a swivel peeler into long ribbons
1 fennel bulb, finely slivered
½ cup (125 mL) white wine
kosher salt to taste
1–2 cups (250–500 mL) veal, beef or brown chicken stock, heated
¼ cup (60 mL) dried cranberries
½ cup (125 mL) whipping cream (optional)
1 Tbsp (15 mL) minced fresh thyme
1 cup (250 mL) finely grated Parmesan cheese
kosher salt and freshly cracked pepper
½ tsp (2.5 mL) grated lemon zest

Put rice and salt into a pot and add four times as much water. Cover and bring to a boil, then reduce heat and simmer for 45 to 60 minutes, until the grains are split open and tender but not mushy. Drain off any excess water, and cover the rice to keep it warm.

In a heavy-bottomed pot, melt butter. Add onion, garlic, bay leaf, rosemary, pepper, zucchini and fennel and cook slowly until the vegetables are tender, stirring gently from time to time. Remove half the vegetables and set aside, covered to keep warm. Add wine to the pot, then stir in stock and cranberries. Bring to a boil and reduce by half, then add the rice and stir gently. Stir in the cream (if using), thyme and grated cheese, salt and pepper. Serve hot, garnished with remaining vegetables and a little grated lemon zest.

on new ground

On a sunny Prairie summer afternoon, clouds are sailing across the vast sky, cumulus ships that coagulate into thunderheads and sudden squalls. From my seat on the house's south-facing deck, I can see the waterfowl that have been attracted to the unexpected lake that arrived with this spring's monumental floods—burnished grebes with masks like carnival revellers, blue-billed ruddy ducks with straight-up tail feathers. Over the rattle of my keyboard, I hear the clatter of the black coots and the hummingbirds' whirring wings.

Beside the deck, nudging up against the gas barbecue, I can see the battered old dairy shipper's milk can my mother painted for me when I opened my restaurant, its logo still perfectly outlined with wheat: "Foodsmith." It's funny how that can has made its way back to its place of origin, just as I have, and how the logo is as relevant now as it was then: a tidy summation of my family name and my lifelong passion for food.

Dave and I have been living here for a year. After twenty-seven years in Calgary, it feels surprisingly like home, this land west of Saskatoon. We have renovated the one-hundred-year-old farmhouse where my grandparents raised my mom and my aunt through the Depression. It's the same house I inhabited for three years with my brothers and my parents. I was a cranky teenager then, and I couldn't wait to shake Saskatchewan's dust off my boots. I left for the bright lights of Vancouver and cooking school as soon as I graduated high school. But here I am, back again, and settled into what feels more homelike than anywhere else.

Buying direct from the producer remains the best way to access the best local food. Although I know Alberta farmers well, relocating meant we had to find "our" Saskatchewan farmers. We go to the farmers' market, we look for farmers who sell direct, at the farmgate or via the Internet, and we search out stores that carry local products.

Instead of joining a community garden or a CSA this spring, we planted a garden. It had its toes in the "one-hundred-and-fifty-year-event" flood-water we live beside, so the asparagus and rhubarb have been suffering, but everything else, including the weeds, is thriving. Beneath the ground, spuds grow their blind eyes; above ground, heirloom tomatoes ripen; green and yellow beans dangle like jewels; leathery beet leaves sway in the breeze. I go out each day in the perfection of the early light to weed, pausing to pick thimble-sized raspberries and snip lettuces for supper. I have resumed my lifelong affair with canning, and fill the basement shelves with gleaming preserves. Our lamb, chickens and pork come from a farmer twenty miles away. Our freezer is filled with raspberries and rhubarb, my home-cured bacon, our neighbours' Highland beef.

Knowing your farmer isn't just about being able to ask how your food is raised. The trade in goodwill between producers and consumers is transferred to the table—goodwill equals good food. It's a triumvirate: the ingredients, the consumers and the growers. Who's my farmer? What does she grow, and how? What yummy dishes can I create with my local harvest? Knowing my farmer re-establishes the all-important tradition of community; relation-ships are the missing ingredient in the modern food chain.

Where we choose to spend our money is a true indicator of what we value. It is illustrative of food's great divide: buy on price or buy on story, shop for a commodity or shop for dinner. Part II of *Foodshed* addresses issues and solutions that are Albertan in their specifics but translate across the country and the continent. Talk to a PEI spud farmer or a peach orchardist in the Similkameen Valley, and their stories eerily echo the tales told by my Albertans. The Appendices offer resources.

When I was younger, I thought that cooking was all about the hands of the maker. But I have learned that a cook is only as good as her ingre-dients, and that the hands of the grower matter just as much. The view from the farmer's field has become my preferred perspective. There is a pleasing sense of rightness that this book was written on the farm where my mother and grandparents lived, in a sunny, second-storey studio that overlooks the south alfalfa field. In the end, when I cook, it's about eating what grows close to home. I wish you the same pleasure.

—The Farmhouse, west of Saskatoon, August 2011

PART TWO
FACTS AND FIGURES

Asparagus viewed from the ground as Doug Edgar's "asparga-buggy" approaches.

changing landscapes, terroir and our global diet

What farmers grow is dictated by locale, or terroir, as the grape-growers succinctly refer to the combination of aspect, sunlight, soil, minerals, water, temperature and heat units that come together to make each vintner's wines unique. Corn in Taber and Shaftesbury; peas and lentils in Bow Island; strawberries in Innisfail. Our language reflects the renewed reality of local eating: locavore; foodshed. Locale is the noun, local is the adjective, and locavore is the creature.

Albertan farmers have already adapted what they plant as weather patterns shift. Our globally sourced diet is bound to change too, as energy and water use alter our food supply and the cost of transportation. A mostly local diet will once again be the reality for most North Americans.

Food is a hot topic around the world, again the primal issue it always has been. In Egypt, the 2011 fall of Hosni Mubarak's thirty-year regime began with marchers who chanted, "Bread and freedom!" In India, eco-feminist, physicist and Slow Food co-founder Vandana Shiva continues to write about the rights of farmers and the health of soil. In the USA, food is front and centre. Books such as *The Omnivore's Dilemma* by Michael Pollan routinely hit the bestseller charts. Health is inextricably linked to poor food habits, as witnessed by flooding epidemics of diabetes and obesity.

sustainability, environmental issues, animals and grass

The Oxford dictionary defines sustainable as that which conserves an ecological balance by avoiding depletion (or contamination) of a species; being caught or farmed in a way that ensures the long-term health and stability of that species, as well as of the greater ecosystem.

Sustainability is a buzzword that has been slapped on everything from houses to hospitality. It is truest and most resonant in an environmental context. It makes perfect sense that those who live close to the land would see and comprehend their utter reliance upon the land's well-being in a way that some urban residents don't. Many farmers and ranchers are born naturalists and environmentalists, and are founding or active members of environmental groups such as the Southern Alberta Land Trust (SALT), the Nature Conservancy, the Weston A. Price Foundation, the Canadian Parks and Wilderness Society, Cows and Fish, and the Alberta Wilderness Association.

In 2008, the Alberta government enacted Bill 36, the Land Use Framework, a controversial land appropriation plan that subdivided the province into seven watershed-based areas, with an aim of developing ". . . healthy economies, healthy eco-systems and environment . . . and people-friendly communities." As Alberta's population and resource needs skyrocket, each of the seven regions must determine its priorities: Urban expansion? Agriculture? Industry and resource exploitation? The watershed and environment?

According to Alberta by Design, a website coordinated by the Canadian Parks and Wilderness Society, the Pembina Institute, and Water Matters, a new vision of land use and stewardship was past due. Oil and gas wells drilled in Alberta have more than doubled from 8,400 in 1997; coal-bed methane wells now number in excess of 12,000; more than 200,000 kilometres of electric transmission lines criss-cross the province. At the same time, nearly one-third of the province's land (over

220,000 square kilometres) is under cultivation, on nearly 50,000 farms and ranches. It makes for an uneasy truce.

Alberta by Design suggests that five issues are on the table: water use; fragmentation and conversion of agricultural land to other uses; financial benefits to growers who maintain sustainable ecosystems; land stewardship tools such as conservation easements (agreements between landowners and conservation groups that limit the amount and type of development on a piece of land); and "cumulative effects measurement" within a region as opposed to project-by-project decision-making.

Overshadowing all is the larger question of who makes the decisions, and what recourse unhappy Albertans have when regional plans don't suit their needs or are implemented through governmental statutory consent. Amendments to the Land Use Framework passed in 2011 include the chance for ". . . persons directly affected by a regional plan to request a review of a regional plan; for title holders to apply for a variance of a plan; for landowners to apply for compensation if they believe their rights have been taken away."

All the farmers, orchardists, market gardeners and ranchers I spoke with and visited for *Foodshed* believe that fewer chemical inputs into the soil are best. Many are reluctant to use the word "organic," and won't jump through the economic and management hoops required to become certified organic. For many of our farmers and ranchers, the health of our land is *the* primary consideration, William Bryant Logan asserts in *Dirt: The Ecstatic Skin of the Earth*.

Land health is at risk from the feedlots where many animals destined for dinner finish their lives. Such spots generate massive amounts of fecal pollution and contaminated runoff water. The provincial government requires feedlots of more than one hundred and fifty animals to apply for a permit, as issued by the Natural Resources Conservation Board (NRCB). Feedlots that generate more than five hundred tonnes of manure are required to comply with specific regulations governing setbacks for manure application, soil testing, application limits and record-keeping. Smaller businesses are expected "to handle their manure in a responsible manner" and the NRCB sends out field staff in response to complaints. In the province's revised Land Use Framework, the province retains control over approval for feedlots, with public input and controls imposed by

Haymows drawn by draft horses can still be found on Albertan farms.

municipalities. I believe Alberta needs a system that is built on more, and smaller, facilities for raising and slaughtering our animals. The resulting shorter trips to the slaughterhouse would have the additional benefit of lessening the animals' stress levels.

Temple Grandin, the autistic animal scientist, was named to *Time Magazine*'s 2010 list of one hundred most influential people. Grandin's research into humane killing and stunning methods has led to widespread reform in animal treatment at abbatoirs. Her curved chutes and corrals reduce the stress of confined cattle, and are in use globally. Growers and consumers are becoming more vocal in their insistence that the animals we consume be raised humanely in uncrowded, natural conditions and be treated kindly until their deaths. As American farmer-poet-philosopher Wendell Berry puts it in his essay "The Pleasures of Eating," published in *The Art of the Commonplace: The Agrarian Essays of Wendell Berry*, ". . . we are living from mystery, from creatures we did not make and powers we cannot comprehend."

In his memoir, *A Hunter's Confession*, David Carpenter, a Saskatchewan writer and former hunter who has renounced hunting, makes a case for reverence toward all animals we eat. In drawing a convincing parallel between how we treat our cultivated-for-food animals and how many hunters view animals, he writes, "What happens when we come to believe that animals are subject of our dominion, merely there for our needs and not there in and for themselves, cohabitants of the planet, so to speak, is that we objectify them without the tiniest regret." Kindness and a respectful, stress-free life and death are necessities for any animal.

In the research for his book *Steak: One Man's Search for the World's Tastiest Piece of Beef*, Canadian writer Mark Schatzker spoke with scientists, ranchers, chefs and meat advocates as he pursued the perfect piece of steak, eating grass-fed and grain-finished beef in Japan, Argentina, Italy, Scotland, France, the USA and Canada. As part of his research, Schatzker raised a trio of cattle, utilizing what he had learned. He accompanied one, Fleurance, to the slaughterhouse before she became part of several meals produced by chef Michael Stadtländer at Eigensinn Farm north of Toronto. He reached the conclusion that grass-fed beef is far superior to feedlot beef—where careful selection of breed, grass and soil are observed.

We have changed our minds about more than the treatment of our animals for humane reasons. We want our food to fuel body and soul. The best fuel, the most nutrient-dense food, comes from animals raised—and finished—on grass, and plants raised without chemicals. According to Schatzker, grass-fed meat contains optimal ratios of omega-3 and omega-6 essential fatty acids. It also contains more conjugated lineolic acid (CLA), which the body cannot produce but which is thought to be an effective anti-cancer agent.

Grass is also better for animals: too much grain can cause acidosis in ruminants, animals who chew their cud. Traditionally, those who graze in summer, and eat hay (dried grass) and silage (fermented grass) in winter do better than those who are fed excessive grain.

government involvement and labour

The Progressive Conservative government of Alberta has sent a mixed and confusing message to consumers and to producers about the value and validity of food, local and otherwise. On the one hand, the government started Dine Alberta in 2002 to bring together consumers, chefs and producers. On the other hand, the government passed Bills 19 and 36, land appropriation laws that do away with citizens' rights while grabbing valuable and irreplaceable farmland for gas plants and mineral extraction. Many of my farmers have struggled with rules and regulations designed for large-scale production. Clearly, a second set of regulations is required for growers seeking to work on a smaller, local scale, along with infrastructure to support local foods systems.

A terrific example of a successful local foods system started by an individual is Local Food Plus, the Ontario-based non-profit organization that journalist Lori Stahlbrand incorporated in 2005. It certifies growers and producers according to rigourous standards that focus on sustainable production, labour, native habitat preservation, animal welfare and on-farm energy use. The second, equally important step links those growers to consumers, chefs and retailers, including McGill University and Toronto's Scarborough Hospital.

Albertan farmers have exhibited ingenuity and sheer tenacity in developing distribution networks, informal marketing co-ops and brokerages to get their food from gate to plate, but it all takes time and energy and diverts farmers from doing what farmers do best. A system similar to Local Food Plus would return farmers to their fields.

Subsidies and quotas are another book entirely, as are oil sands issues that are linked to food and the environment. But many looming questions—access to land, land appropriation, making a living, planning for the future—are addressed in *Foodshed*'s pages by our growers, in their own voices. The same questions are being raised in other communities around the globe.

Immigrant labour built Canada, and there is nothing pretty about the exploitative side of the equation. First, the railroads, their tracks laid by Chinese labourers; now, in our agriculture industry, our slaughterhouses are predominantly staffed by Sudanese, and many of our farms employ seasonal Mexican, Thai and Philippine workers who leave their families behind. They do the hard physical work that our Canadian sons and daughters don't want to perform for low wages.

We do not all have to return to a rural lifestyle and grow our own food. But the labour solution *is* societal, rooted in how we regard farmers, and the low priority we accord to our food. According to the United States Department of Agriculture, and Agriculture and Agri-Food Canada, Canadians are on a fifty-year downward spiral in food spending, from nearly 20 per cent of family budgets to less than 10 per cent—half of what western Europeans spend. We spend our money on what we value, and change happens when the need becomes personal. Would consumers eat better if they had to pay a tax on junk food and received a tax credit for buying food from its source?

The bottom line, as ever, is education. Organizations like Slow Food provide that contact, as do CSAs and farmers' markets. Urban dwellers need access to farms, and their farmers, to learn how our food is raised, to understand the attraction of the hard but rewarding work it engenders, and to simply put foot on the land.

finding local food

It's easy to shop in large supermarkets. But dissatisfaction with the choices they offer has led to a boom in farmers' markets, farmgate sales, CSAs and community gardens.

Urban Growers

For some of us, the grower is the face in the mirror. Community and urban gardens are visible in Alberta as never before, and are documented in my colleague Jennifer Cockrall-King's new book, *Food and the City: Urban Agriculture and the New Food Revolution.* In Calgary, the Community Garden Resource Network, a three-year project of the Calgary Horticultural Society, traces urban gardens to 1914 with the Vacant Lots Garden Club. In 2009, twenty-one public community gardens and twenty-five private community gardens were created, up from a total of eleven gardens in 2008. The "geometric progression" of 2010 amounted to over sixty gardens. The Community Gardens Network of Edmonton lists seventy community gardens, an interesting reversion to the staggeringly high number of gardens tended by immigrants along the North Saskatchewan River in the city's early days.

The Canadian Pacific Railroad, which built the continent's first transcontinental railroad to bring settlers west, also played a large role in Canadian farm-and-garden lore between 1890 and 1930. According to railroad historian Oana Capota, by 1912, over fifteen hundred flower and vegetable gardens were attached to many western Canadian train stations. They were part of a well-thought-out marketing plan, but were also a sustainably designed loop: steam engines' steam was discharged as water for the gardens in barrels along the train tracks, and some of the greenhouses were heated by laundry steam. The vegetables fed travellers and station staff while serving as a barometer of the potential beauty and fertility of the rough new land that faced immigrants and settlers. Survival, then as now, depended on cultivating good gardens. In the years of the First and Second World Wars, Relief and Victory Gardens provided food for the war efforts.

Farmers' Markets, Farmgate Sales, Direct Sales

The Alberta Farmers' Market Association reports the existence of over one hundred government-approved farmers' markets province-wide, involving over three thousand vendors. Alberta's oldest is Edmonton's City Market—in existence since 1900—and a baker's dozen markets feed Edmontonians, plus St. Albert's, the largest outdoor farmers' market in western Canada. A single market is located in Lethbridge. In 2001, there were six farmers' markets in Calgary; in 2010, I observed a record number within the city's limits—ten farmers' markets plus three public markets. The differentiation is a big one: a true farmers' market is one where vendors make, bake or grow what they sell.

Other farmers and growers have opted, in growing numbers, to sell direct to consumers—from the farmgate, or through online pre-ordering/delivery systems.

CSAs

Community Supported Agriculture (CSA) is a community of individuals (or families) who support a chosen farm and its family. Each purchases a share of the year's crop in advance. Thus, the customers become virtual partners (or "co-producers," in Slow Food founder Carlo Petrini's words) in the farm, sharing risk and reward with the farm family. Throughout the growing season, usually sixteen weeks long, each subscribing member receives a weekly bag, box or basket of freshly harvested food from the farmer.

CSAs have their philosophical roots in the writings of Austrian philosopher Rudolph Steiner. Steiner formulated his educational and agricultural theories in the 1920s. Primary among them was the association of producers and consumers, where consumer and producer are linked by their mutual interests. Ideally, this would lead to an economy where what is produced locally is also consumed locally.

The actual farms originated in a number of regions around the globe. Consumers and farmers formed community farms in Chile under the regime of Salvatore Allende in the early 1970s. The Chilean co-op movement inspired a Swiss farmer to start a community-based farm, and in 1960s Germany, consumers versed in Steiner's work were interested in founding an agriculture system that was ecologically sound and socially equitable. In Japan, meanwhile, mothers concerned about pesticide usage, the increase in imported food and loss of arable land to encroachment by urban growth began similar subscription-funded farming called *teikei*.

CSAs focus on the production of high-quality foods for a local community. Consumer involvement results in a stronger consumer-producer relationship and can lead to the formation of a cohesive community. The more a farm embraces "whole-farm, whole-budget support," the more it can focus on quality and reduce the risk of food waste or financial loss. It's a win-win arrangement: consumers gain a community and a relationship with the farmer who feeds them. The consumer knows how her food is grown, and what, if any, substances are used in its growth. A CSA subscriber may visit the farm and get some sunshine and soil on her hands, in addition to seasonal, fresh food that has not travelled very far. A CSA member regains a measure of control over her local food supply. Learn how CSAs operate in the Appendices.

Slow Food

Slow Food has had an impact on Albertan food since its arrival in 2000. "Local" is Slow Food's strength. Individual Slow Food *convivia*, or chapters, hold events that focus on local products and producers, connect consumers with their farmers, and promote taste educations to re-awaken and train the senses. Read more in the Appendices.

home cooking

All this talk about ingredients! It presupposes that most of us cook at home. I'd like to report that more people are actually cooking at home, but a simple trip to the market shows otherwise, despite a plethora of televison food shows. Stores are filled with ready-to-heat dishes that require no cooking, and kids are not learning how to cook at home. My years of teaching have shown me that simply watching food telelvsion is not enough: information is not knowledge. Cooking is a databank of tactile and sensory skills, skills learned by doing, not by passively watching. And much as I applaud television for popularizing food, it's the next step that counts most: turning on the stove and setting a pan on the flame.

I have heard stories from other educators and from market gardeners who see first-hand that their visitors and charges do not know how their food grows. Chef Andrew Hewson, a culinary educator at SAIT Polytechnic in Calgary, was appalled one day when he sent a student to the cooler for a bunch of basil. The kid returned clutching a handful of parsley. We have ourselves to blame: packaged food is not real food.

The late scholar and theologian Thomas Berry says this in his 2000 book, *The Great Work: Our Way into the Future*: "Our children no longer learn how to read from the great Book of Nature from their own direct experience or how to interact creatively with the seasonal transformations of the planet. They seldom learn where their water comes from or where it goes. We no longer coordinate our human celebration with the great liturgy of the heavens."

The good news is that cooking classes are growing easier to find. Returning to the kitchen stove returns control over what we eat to us. It's not just that home cooking tastes better and re-opens the channel to shared mealtimes and the sensory enjoyment of food, but it also pulls the teeth out of fretting about additives, salt content, and the terrible things masquerading as food on supermarket shelves.

Living more simply, requiring less of our world, eating more season-ally, buying our food from people we know, and setting our feet on the soil outside of cities are healing things. It's not just so our children know where and how their food is raised. Standing on a piece of fertile and growing land is soul-renewing.

Appendix A:

CSAs

CSAs spread to North America in 1986 with two farms, one in Oregon, the other in New Hampshire, and spread inland. According to the USDA's 2007 agriculture census report, released in 2009, 12,549 CSAs operate in the USA, although exact numbers are hard to establish as many CSAs maintain a low profile.

CSAs by definition are locally based, and remain a grassroots approach to farming, with no cohesive structure or national database. However, nearly every province in Canada has some sort of online CSA listing. As the Ontario site's webmaster confirms, although the site lists more than 150 Ontario CSAs, there are many other CSA operators in the province who do not know about the site and are not listed. According to Equiterre, the Quebec-based non-profit organization that works on social and environmental issues, 8,300 Quebec families receive weekly baskets from 78 family farms.

BC's FarmFolkCityFolk, a non-profit organization committed since 1993 to supporting a local sustainable food system, lists 34 CSAs on its website. Upwards of six CSAs are in existence in Saskatchewan, according to the Saskatchewan Organic Directorate (SOD). Atlantic Canada Organic Regional Network (ACORN), a non-profit organization founded in 2000, lists fewer than ten CSAs on its website. In Alberta, at last count, 16 farms offered CSA subscriptions, a major increase from three operating CSAs in 2008. The oldest, Tipi Creek, began in 1994. An additional six farms or individuals offer variants.

Many CSAs practise organic (certified or not) or holistic ecologically aware tillage methods. In most cases, animals are free range or pastured. Two Albertan CSAs use animals—draft horses or oxen—to provide power instead of tractors, insisting that animals tread more lightly on the soil.

Membership Rates and Variants

Farmers sell annual subscriptions to a set number of clients before the onset of spring planting, usually giving first refusal to the previous year's clients before new clients are solicited. Full shares are suitable for a family of four. Some farms offer half shares, suitable for two to three people, for slightly more money than half the cost of a full share.

Some CSAs charge extra for options such as eggs or meat. A reduced-rate work-share membership is offered to those who wish to regularly work on the farm. Most CSAs require payment in full in March or earlier, to cover costs and provide farmers with income in the off-season; others allow subscribers to divide their fee into several payments over the course of the year or the season.

Seasonality

What a farm grows is dictated by geography, terroir and a farmer's preferences. Most Albertan CSAs offer vegetables and fruit, usually all grown on the farm, although some act as co-operatives, bringing together a variety of foodstuffs from multiple farms so that their subscribers have access to a wider array of foods over a longer season. Homegrown meat, cheese, eggs, dairy products, grain, flowers and pulses supplement their vegetables and/or extend the season into wintertime. Some farmers preserve some of their goods by freezing, canning or drying, for distribution in later months. The occasional CSA specializes, offering only cheese, meats, dairy or grain.

Delivery and Communication

Most farms organize one or more central drop-off spot(s), often at a farmers' market in a nearby town or city where subscribers gather weekly to collect their share of the week's yield. Some farmers institute strict pickup times and shares left uncollected after a set time are sold to the public, given to organizations that feed the disadvantaged or divided among other members. A few farms require subscribers to come to the farm to pick up.

Many CSAs have websites, blogs or weekly e-mails to subscribers. They write about what is in season each week, include recipes, storage tips for produce, stories about the farm and its residents, and news of current activities and the farm's state of weather and other affairs.

Frequently, during pickups, subscribers informally exchange recipes and cooking tips, and often establish relationships with the other subscribers. At pickups, during the small talk that ensues while waiting for their goods, subscribers learn who lives where, and some set up carpools or take turns collecting the week's goods.

Some CSA arrangements require on-farm work days. Seeding, weeding and hoeing, watering and harvesting, fencing, building and maintaining structures like greenhouses and packing sheds, and end of season cleanup are some of the ways subscribers provide on-farm assistance. Others offer an annual optional "farm visit day" to their subscribers, with or without a labour component, usually including a meal. Both options give urban residents a chance to see the garden, the animals, the fences and the farm itself. Some offer lessons in animal husbandry, driving draft animals, preserving and harvesting.

Variations

Some CSAs supply some fraction of their goods to buying clubs or brown box programs. Such clubs or co-ops typically operate in one location, often a house or the back end of a member's business. They acquire local produce and products for members, and usually charge a membership fee in addition to whatever members order. Ordering is often done online or by e-mail. Another variant is a direct sales version, where customers are not subscribers, but pre-order online or via e-mail on a regular schedule (weekly, biweekly) and pick up their orders at a preset central location.

Advantages

CSAs benefit consumers and farmers.
- They simplify a farmer's life. They reduce the need for farmers to spend a lot of time, effort and expense learning how to market, package, label and distribute their food. At the same time, CSAs eliminate the time and risk inherent in attending farmers' markets. Having a guaranteed market means knowing exactly how much produce to harvest and package each week. A CSA farmer knows who is eating the food grown on her land. This intangible capital forms the basis of community.
- Funds are provided by subscribers at the beginning of the season, up-front, when the farmer needs cash to buy seeds, maintain or build structures, repair equipment, and to cover living costs in the off-season.
- In Alberta, food produced on CSA farms is consumed on average within 120 kilometres of where it was grown, requiring a minimum of transportation costs.
- Most CSAs are organic, natural or biodynamic so the chemical input load on local land is reduced.

- CSAs keep farmers on the land, growing food for local consumers. They are an integral part of a local food system.
- CSAs offer opportunities for learning new skills—food preservation, animal husbandry, care and usage of draft animals, sustainable gardening.

Disadvantages

- CSAs must be close to a large population base to attract clients. As a result, CSAs are mostly situated in densely populated parts of the province, where agricultural land is at a premium.
- When a particular food is in season, it's in season in abundance. When the season concludes, it's done. No asparagus is available after late June, but in July through September, subscribers may have more Swiss chard than they appreciate. Some CSAs turn this into a plus for their other members by setting up a Swap Box.
- Weather is the always-present wild card. If the weather is unkind or untimely, crops may fail to germinate, fail to set blossom, fail to grow or fail to set fruit. Plants may freeze, drown or dehydrate. Chickens may stop laying, cows may go dry, lambs may not thrive. That is the risk the farmer has always borne, so the CSA subscriber bears it too, and gets as much or as little as Mother Nature provides.
- Subscribers must in most cases put up cash money in a large payment "up front."
- The week's goods must be cleaned and divided into as many bags or boxes as there are shares. This may entail initiating a bag-recycling program with subscribers, or purchasing twice the number of boxes as subscribers so that one is always at the farm for filling.
- Planning on-farm events and subscriber work days takes a lot of planning and energy to organize. Any value-added processing must be done in an approved and certified commercial kitchen.
- Average weekly CSA costs to subscribers may be more or less than a family's pre-existing supermarket produce budget, depending on their shopping habits. For many, knowing the source and practices of the grower outweighs any cost differential.
- CSA income for farmers is variable, and in many cases insufficient to support a family. Many hold an off-farm job to supplement their farming income.

Appendix B:

Organics

On June 30, 2009, Canada's Organic Products Regulations (OPR) came into effect, making the new Canadian Organic Standards (COS) mandatory. The regulations require that any organic products must be COS-certified in order to cross provincial or international borders or use the Canada Organic Logo.

This is an attempt to solve the varying requirements that have arisen among the certifying bodies. Organics in the modern era (post-Second World War) was a grassroots phenomenon that followed the 1960s hippies and back-to-the-landers, so certifying bodies tended to evolve where growers congregated, usually with little correspondence among them.

Organic farmers produce an audit trail. Its purpose is to trace the product from the farm to the processor to the consumer, and provides provenance to processors, retailers and consumers. It is the producer's proof that they have used acceptable organic growing practices in their production system. The audit trail includes a farm map, field history records, input records, harvest records (location, product and date), storage records/bin inventory, sales records (including sale date, commodity sold, lot number and amount, as well as transaction certificates used for all sales).

Organic certification guarantees production based on organic standards. Certification is an annual third-party inspection to confirm that the crops, animals, process or service conforms to organic standards. The standards deal as well with storage, transportation, packaging and processing of organic products. Private certification bodies that provide organic certification set varying prices and have their own compliance requirements. Where a producer intends to market can affect which certifying body is chosen. The certification process includes a review of documentation, on-site inspection and assessment.

Requirements for starting materials used for production vary with differing certifying bodies. Does the planted seed need to be certified organic? Do the mothers of organic offspring need to be certified organic?

Processors must adhere to the standards of certification bodies as well.

These standards apply to transportation and storage, processing, packaging and labelling for processor certification.

Regulations include:
- Organic products cannot be mixed with non-organic products during storage and transportation.
- Processing of organic animals must take place in certified facilities.
- Only permitted food additives and processing aids can be combined with the product during processing.
- Only permitted pest control can be used on or near the product.
- A processed product can be labelled as organic if at least 95 per cent of the ingredients, excluding added water or salt, are obtained from certified organic sources.

Seven Principles of Organic Farming
- Protect the environment and promote a sound state of health.
- Maintain long-term soil fertility.
- Maintain biological diversity.
- Recycle materials and resources.
- Promote livestock health.
- Maintain organic integrity.
- Rely on renewable resources in locally organized agricultural systems.

Conventional foods that several different studies have found to have the highest pesticide residues include apples, butter, celery, cherries, cucumbers, grapes, green beans, milk and milk products, peaches, peanuts, pears, popcorn, potatoes, spinach, strawberries, bell peppers and squash.

Appendix C:

Slow Food

> "Slow Food unites the pleasure of food with responsibility, sustainability and harmony with nature."
>
> —Carlo Petrini

Carlo Petrini, an Italian journalist and co-founder of Slow Food, calls consumers "co-producers" because of their necessary role in the circle of food production and consumption. Petrini founded Slow Food in 1989, in outraged response to two events: the arrival of McDonald's "golden arches" adjacent to the Spanish Steps in Rome; and attending an annual tomato festival where he learned that farmers could not access seeds for certain heirloom varieties because of the control multinational corporations had over seeds.

Slow Food International today has over 100,000 members, in 150 countries, who share Petrini's belief that slow is better than fast food or fast living, and that the health of the planet is related to the health of what is on our plates. But Slow Foodies have a sense of humour— the international movement's symbol is the snail, which calmly glides through life, eating as it goes.

Slow Food focuses on five distinct areas: preserving traditional methods of growing and consuming foods; the pleasures of the table; promoting networking between consumers, chefs and producers; defending biodiversity; and promoting a thriving local economy.

Restoring the cultural dignity of food is the goal at the University for Gastronomic Sciences in Pollenzo and Colorno, Italy. This unique place supports high-level research and education by international experts.

Slow Food's Foundation for Biodiversity records plant species and animal breeds at risk of extinction, and inducts them into a metaphorical Ark of Taste. The Ark is more than a notion; it takes action by managing Presidia, projects that safeguard animal breeds and plant varieties, protect traditional production methods that have cultural and economic relevance, and save outstanding food products in their place of origin, particularly in developing countries. Thus, the Ark's influence can save a species or product, but also the human lives that depend on it. Currently, among the hundreds of Ark spots occupied by global foods, there are seven Canadian nominees: Nova Scotia's Gravenstein apple; herring spawn on kelp; Canadienne cow; Chantecler chicken; Montreal melon; great plains bison; and Red Fife wheat.

The Slow Food Foundation for Biodiversity also organizes Terra Madre, Slow Food's biennial conference where thousands of farmers and food producers from more than 120 nations meet. Slow Food convivia around the world nominate cooks/chefs and producers who grow, raise, catch, distribute and create food in ways that respect the environment, defend human dignity and protect the health of consumers. Three of Alberta's four convivia are ironically located in the two major cities. Nearly 200 Albertan delegates have gone to the three Terra Madre gatherings held since its inception in 2006.

Slow Food is really slow living. The movement has spawned *Cittaslow*, or Slow Cities, a global group of towns and cities committed to improving their citizens' quality of life by making their cities more pleasant places to dwell. This could mean closing the city core to auto traffic for a day per week, or creating opportunities for producers and consumers to interact. Cowichan Bay, on Vancouver Island, and Naramata, in the Okanagan, are Canada's only Slow Cities.

Underscoring all is a belief in the necessity of "good, clean and fair" food. The food we eat should taste good; it should be produced in a clean way that does not harm the environment, animal welfare or our health; and food producers should receive fair compensation for their work.

Appendix D:

Resources: Books and DVDS

100 Mile Diet: A Year of Local Eating, The by Alisa Smith and JB MacKinnon (Vintage)

Animal Vegetable Miracle: A Year of Food Life by Barbara Kingsolver (HarperCollins)

Anita Stewart's Canada by Anita Stewart (HarperCollins)

Apples to Oysters: A Food Lover's Tour of Canadian Farms by Margaret Webb (Viking)

Art of Simple Food, The by Alice Waters (Potter)

Art of the Commonplace: The Agrarian Essays of Wendell Berry, The (Counterpoint)

Basic Formula to Create Community Supported Agriculture by Robin Van En (Indian Line Farm, Box 57, Jugend Road, Great Barrington, MA 02130)

Biopiracy: The Plunder of Nature and Meaning by Vandana Shiva (South End Press)

Cod: A Bibliography of the Fish That Changed the World by Mark Kurlansky (Random House)

Coming Home To Eat: The Pleasures and Politics of Local Foods by Gary Paul Nabhan (Norton)

Community Supported Agriculture (CSA): Making the Connection by Bill Blake *et. al.* (UC Cooperative Extension, Attn: CSA Handbook, 11477 E Avenue, Auburn, CA 95603)

Cooking by Hand by Paul Bertolli (Potter)

Dirt: The Ecstatic Skin of the Earth by William Bryant Logan (Riverhead Books)

Edible Journey: Exploring the Islands' Fine Food, Farms and Vineyards, An by Elizabeth Levinson (TouchWood Editions)

End of Food: How the Food Industry is Destroying Our Food Supply and What We Can Do About It, The by Thomas Pawlick (Greystone)

Epitaph for a Peach: Four Seasons on my Family Farm by David Mas Masumoto (HarperCollins)

Ethical Gourmet, The by Jay Weinstein (Broadway)

Everything I Want to Do is Illegal by Joel Salatin (Acres USA)

Farmageddon by Brewster Kneed (New Society Publishers)

Farmer John's Cookbook: The Real Dirt on Vegetables by Farmer John Peterson and Angelic Organics (Gibbs Smith)

Farms of Tomorrow Revisited: Community Supported Farms—Farm Supported Communities by Trauger Groh and Steven McFadden (Biodynamic Farming & Gardening Association)

Fast Food Nation: The Dark Side of the All-American Meal by Eric Schlosser (Houghton Mifflin)

Feeding the Future: From Fat to Famine, How to Survive the World's Food Crises eds. Andrew Heintzman and Evan Solomon (Anansi)

Fields of Plenty by Michael Ableman (Raincoast)

Food & the City: Urban Agriculture and the New Food Revolution by Jennifer Cockrall-King (Prometheus Books)

Food Lover's Trail Guide to Alberta, The, Volumes I & II by Mary Bailey and Judy Schultz (Blue Couch Books)

Food Politics: How the Food Industry Influences Nutrition and Health by Marion Nestlé (University of California Press)

For Hunger-Proof Cities: Sustainable Urban Food Systems eds. Mustafa Koc, Rod MacRae, Luc J.A. Mougeot and Jennifer Welsh (International Development Research Center)

Fresh: Seasonal Recipes Made with Local Foods by John Bishop and Dennis Green with Dawn Gourley (Douglas & McIntyre)

From the Good Earth by Michael Ableman (Harry N. Abrams)

Genetic Roulette: The Documented Health Risks of Genetically Engineered Foods by Jeffrey M. Smith (Yes! Books)

Grass, Sky, Song: Promise and Peril in the World of Grassland Birds by Trevor Herriot (HarperCollins)

Hunter's Confession, A by David Carpenter (Greystone)

In Defense of Food: An Eater's Manifesto by Michael Pollan (Penguin)

In Praise of Slow: How a World-wide Movement is Challenging the Cult of Speed by Carl Honoré (Knopf)

Invisible Giant: Cargill and its Transnational Strategies by Brewster Kneen (Pluto Press)

Local Flavors by Deborah Madison (Broadway)

Locavore: From Farmers' Fields to Rooftop Gardens—How Canadians are Changing the Way We Eat by Sara Elton (HarperCollins)

Menus From an Orchard Table: Celebrating the Food and Wine of the Okanagan by Heidi Noble (Whitecap)

Nourishing Traditions by Sally Fallon with Mary G. Enig (New Trends Publishing)

Oldways Table: Essays & Recipes From the Culinary Think Tank, The by K. Dun Gifford and Sara Baer-Sinnott (Ten Speed Press)

Omnivore's Dilemma: A Natural History of Four Meals, The by Michael Pollan (Penguin)

On Food and Cooking: The Science and Lore of the Kitchen, 2nd Ed. by Harold McGee (Scribner)

On Good Land by Michael Ableman (Raincoast)

Our Field: A Manual for Community Shared Agriculture by Tamsyn Rowley and Chris Beeman (University of Guelph, Guelph, Ontario)

Physiology of Taste or Meditations on Transcendental Gastronomy, The by Jean Anthelme Brillat-Savarin, translated by MFK Fisher (Counterpoint)

Pleasures of Slow Food: Celebrating Authentic Traditions, The by Corby Kummer (Chronicle Books)

Prairie Feast: A Writer's Journey Home for Dinner by Amy Jo Ehman (Coteau Books)

Sharing the Harvest: A Citizen's Guide to Community Supported Agriculture by Elizabeth Henderson, with Robyn Van En (Chelsea Green Publishing Company)

Sheer Ecstasy of Being a Lunatic Farmer, The by Joel Salatin (Acres USA)

Slow Food Nation's Come to the Table: The Slow Way of Living ed. Katrina Heron (Rodale)

Slow Food Revolution: A New Culture for Eating and Living by Carlo Petrini (Rizzoli)

Slow Food Story: Politics and Pleasure, The by Geoff Andrews (McGill-Queen's University Press)

Soil Not Oil: Environmental Justice in an Age of Climate Crisis by Vandana Shiva (South End Press)

Steak: One Man's Search for the World's Tastiest Piece of Beef by Mark Schatzker (Viking)

Stuffed and Starved: Markets Power and the Hidden Battle for the World's Food System by Raj Patel (HarperCollins)

Sustainable Kitchen, The by Stu Stein (New Society Publishers)

Trauma Farm by Brian Brett (Douglas & McIntyre)

Way We Eat: Why Our Food Choices Matter, The by Peter Singer (Rodale)

DVD: *Food Inc.* directed by Robert Penner (available online, at libraries and in stores)

DVD: *It's Not Just About Vegetables* co-produced by Mickey Friedman, John MacGruer, and Jan VanderTuin (available online, at libraries and in stores)

DVD: *Real Dirt on Farmer John, The* directed by Taggart Siegel (available at libraries and in stores)

Appendix E:

Resources: Related Websites

www.acornorganic.org

www.acresusa.com

www.agric.gov.ab.ca

www.albertafarmfresh.com

www.angelic-organics.com

www.biodynamics.com

www.brazeaukitchenparty.ca

www.cban.ca

www.cittaslow.org

www.cog.ca

www.cowsandfish.org

www.csaalberta.com

www.csafarms.ca

www.cuisinecanada.ca

www.curiouscook.net

www.earthtimes.org

www.eatwild.com

www.equiterre.org

www.ffcf.bc.ca

www.foodgirl.ca

www.foodsecurecanada.org

www.foodsecurityalberta.ca

www.heritageharvestseed.com

www.itsnotjustaboutvegetables.com

www.localfoodplus.ca

www.organic.org

www.organicalberta.org

www.organicfarmdirectory.ca

www.reallygoodwriter.com

www.slowfood.ca

www.slowfood.com

www.slowfoodcalgary.ca

www.slowfoodedmonton.ca

www.templegrandin.com

www.trevorherriot.blogspot.com

Appendix F:

Maps, Growers and Farmers: contact information

A is for asparagus

Edgar Farms, Innisfail
  Elna and Doug Edgar, Keri and Randy Graham
  403-350-0659; 403-227-2443
  www.edgarfarms.com;
  www.innisfailgrowers.com

B is for berries

Kayben Farms, Okotoks
  Judy, Jolene, Stephanie, Alexis and Claude Kolk
  403-938-2857
  www.kayben.com

Birds & Bees Organic Winery and Meadery
  (formerly En Santé Organic Winery &
  Meadery), Brosseau
  Elizabeth, Xina and Tonia Chrapko
  780-657-2275
  www.birdsandbeeswinery.com

Field Stone Fruit Wines and Bumbleberry
  Orchards Inc., Strathmore
  Linden, Elaine and Marvin Gill,
  Lorraine and Glen Ellingson
  403-934-2749
  www.fieldstonefruitwines.com

Bridgeview Gardens, Peace River
  Dan and Gail Marusiak,
  Mike and Sheila Marusiak
  780-624-1335
  no website

C is for chicken (and turkey)

Country Lane Farms Ltd., Strathmore
  Jerry and Nancy Kamphuis
  403-934-2755
  www.countrylanefarms.com

Sunworks Farm, Armena
  Ron and Sheila Hamilton
  1-877-393-3133; 780-672-9799
  www.sunworksfarm.com

Sunshine Organic Farm, Warburg
  Ed and Sherry Horvath
  780-848-2288
  www.sunshineorganicfarm.com

Harmony's Way Farm, Crooked Creek
  Larry and Sue King
  780-957-2115
  www.harmonyswayfarm.com

Winter's Turkeys, Dalemead
  Corinne Dahm and Darrel Winter
  403-936-5586
  www.wintersturkeys.ca

D is for dandelions and other greens

Inspired Market Gardens, Edmonton
  Gwen Simpson
  780-910-9363
  www.inspiredgardens.ca

E is for elk

Elbow Falls Wapiti, Priddis
  Win Niebler
  403-931-2427
  www.albertaelk.com

Canadian Rocky Mountain Ranch, DeWinton
  Dr. Terry Church
  403-256-1350
  www.crmr.com

F is for fish

Cunningham's Scotch Cold Smoking, Pincher Creek
  Joe Cunningham
  403-627-6594
  www.joelreadcunningham.com

Greenview Aqua-Farm, Delacour
  Yan Qian
  403-285-3333
  no website

Swift Aquaculture, Ponoka
  Bruce and MaryLou Swift
  604-796-3497
  no website

Headwater Fisheries Inc., Medicine Hat
  Kellen Wickenheiser
  403-952-9097
  no website

G is for grass-fed beef

Trail's End Ranch, Nanton
Linda Loree, Rachel and Tyler Herbert
403-646-3257; 403-646-2550
www.trailsendbeef.com

Sun to Earth Farm, Castor
Richard Griebel and Kathleen Charpentier
403-882-3935
no website

Hoven Farms, Eckville
Tim and Lori Hoven
403-217-2343; 403-302-2748
www.hovenfarms.com

The Producers of the Diamond Willow Range &
Diamond Willow Organics Ltd., Pincher Creek
Keith and Bev Everts
403-627-1800
www.diamondwillow.ca

TK Ranch, Hanna
Dylan and Colleen Biggs
403-578-2404; 1-888-857-2624
www.cattle-handling.com, www.tkranch.com

H is for honey

Kemp Honey, High Prairie
Rachel and Ryan Kemp
780-523-5657
www.kemphoney.ca

Chinook Honey Co., Chinook Arch Meadery and
Chinook Vinegar Works, Okotoks
Cherie and Art Andrews, Pamela Vipond
403-995-0830
www.chinookhoney.com

Lola Canola Honey, Bon Accord
Patty Milligan
780-921-3657; 1-877-921-3657
www.lolacanola.com

I is for iceberg and other lettuces

The Jungle, Innisfail
Leona and Blaine Staples
403-227-4231
www.innisfailgrowers.com;
www.thejunglefarm.com

J is for jalapeno and other chile peppers

Broxburn Vegetables & Café, Lethbridge
Paul and Hilda de Jonge
403-327-0909
www.broxburn-vegetables.com

Oxyoke Farms, Linden
Robby and Phyllis Fyn
403-546-0110
www.oxyokefarms.com

Gull Valley Greenhouse, Gull Lake
Phil Tiemstra
403-885-2242
www.gullvalley.ca

K is for kale

Thompson Small Farm, Sundre
Jonathan Wright and Andrea Thompson
403-638-5787
www.thompsonsmallfarm.ca
www.newfarmer.ca

Sparrow's Nest Organics, Opal
Graham Sparrow
780-942-2259; 780-267-2259
www.sparroworganics.com

L is for lamb

Driview Farms, Fort Macleod
Gerrit and Janet van Hierden
403-553-2178
www.driviewfarms.ca

Cakadu Heritage Lamb, Innisfail
Denis and Linda Jabs
403-728-2398
no website

Ewe-Nique Farms, Champion
Bert and Caroline Vande Bruinhorst
403-897-3737; 403-894-5437
www.eweniquefarms.com

M is for "milk's immortal leap": cheese

The Cheesiry, Kitscoty
Rhonda Zuk Headon
780-846-2590; 780-522-8784
www.osolmeatos.com

Sylvan Star Cheese Farm, Red Deer
Jan, Jannie and Jeroen Schalkwijk
403-340-1560
www.sylvanstarcheesefarm.ca

Smoky Valley Goat Cheese, Smoky Lake
Holly and Larry Gale
780-383-3798
www.smokyvalleygoatcheese.com

Old West Ranch Ltd., Mountain View
James and Debbie Meservy
403-634-7233; 403-653-2331
www.oldwestranch.ca

Fairwinds Farm, Fort Macleod
Ben and Anita Oudshoorn
403-553-0127; 403-553-0037
www.fairwindsfarm.ca

N is for navy beans, (great) northern beans, and other dry beans, lentils and pulses

Owen Cleland, Bow Island
403-223-2772
no website

Leffer Farm, Coaldale
Cornelius, Howard and Monique Leffers
403-345-4885
no website

Saunders Farms Ltd., Taber
Jason Saunders
403-382-1831
no webiste

O is for oilseeds

Highwood Crossing Farm Ltd., Aldersyde
Tony and Penny Marshall
403-652-1910
www.highwoodcrossing.com

P is for pork

First Nature Farms, Goodfare
Jerry Kitt
780-356-2239
www.firstnaturefarms.ab.ca

Sunrise Farm, Killam
Don and Marie Ruzicka
780-385-2474
www.sunrisefarm.ca

Broek Pork Acres, Coalhurst
Allen and Joanne Vanden Broek
403-381-4753
www.broekporkacres.com

Q is for quackers (ducks)

Noble Duck Farms, Nobleford
Gerwin and Esther Van Deuveren
403-315-9510
no website

Greens, Eggs & Ham, Leduc
Andreas and Mary Ellen Grueneberg
780-986-8680
www.greenseggsandham.com

R is for roots

New Oxley Ranch, Claresholm
Jackie Chalmers
403-625-5270
www.newoxley.com

Eagle Creek Farms Inc., Bowden SunMaze and
Eagle Creek Seed Potatoes, Bowden
John and Rayell Mills
403-224-3993; 1-877-224-3939
www.sunmaze.ca; www.eaglecreekfarms.ca;
www.seedpotatoes.ca

Poplar Bluff Farm, Strathmore
Rosemary Wotske
403-934-5400
www.poplarblufforganics.com

Lund's Organic Farm, Innisfail
Gert and Betty Lund
403-227-2693
www.lundsorganic.com

Red Willow Gardens, Beaverlodge
Eric and Carmen Deschipper
780-354-8211
no website

S is for squash

Prairie Gardens Adventure Farm, Bon Accord
Tam Anderson
780-921-2272; 1-888-921-2270
www.prairiegardens.org

Tipi Creek Farm, Morinville
Ron and Yolande Stark
780-459-8390
www.tipicreek.ca

Linda's Market Gardens Ltd., Smoky Lake
Linda and Don Christensen
780-656-2401
www.lindasmarketgarden.com

The Garden of Van Ee-den, Rosemary
Aaron and Barb Van-Ee
403-378-4420
www.vaneedenupick.com

T is for tomatoes

Hotchkiss Herbs & Produce, Rocky View
Paul and Tracy Hotchkiss
587-952-0086
www.hotchkissproduce.com

Paradise Hill Farm, Nanton
Tony and Karen Legault
403-646-3276
www.paradisehillfarm.ca

Hillside Greenhouses, Bowden
Carmen and José Fuentes
403-224-2841
www.innisfailgrowers.com

U is for u-pick

Sprout Farms Apple Orchards, Bon Accord
Amanda Chedzoy
780-921-3460; 1-800-676-0353; 780-910-8847
www.sproutfarms.ca

Antelope Creek Road Berry Farm, Brooks
David and Elizabeth Houseman
403-362-6691; 403-501-4069
no website

The Blooming Fields, Didsbury
Mary-Ann and Pim van Oeveren
403-335-8264; 403-559-9280
www.thebloomingfields.com

V is for vegetables

Blue Mountain Biodynamic Farms, Carstairs
Kristian Vester and Tamara Brunt
403-337-3321
www.bluemountainbiodynamicfarms.com

Nature's Way Veggie Patch, Peace River
Lisa, Peter and Mary Lundgard
780-835-4685; 780-338-2934
www.buylocalthinkglobal.com/organic

Dunvegan Gardens, Dunvegan, Edmonton and
Fort McMurray
Brad, Ron and Pauline Friesen
780-835-4459; 780-791-4363; 780-791-0212
www.dunvegangardens.com

Jensen Farms, Taber
Allen, David and Susan Jensen
403-223-8385
www.jensenstabercorn.com

W is for wheat

Heritage Harvest, Strathmore
Henry Winnicki, Mark Gibeau, and
Ray LeFebvre
Mark Gibeau: 403-934-3457
Ray LeFebvre: 403-934-5809
no website

Ehnes Organic Seed Cleaning Ltd. and Back 40
Organics Ltd., Etzikom
Bernie Ehnes
403-666-2047
www.ehnesorganic.com

X is for xeriscaping experts, nearly extinct (plains bison)

Olson's High Country Bison, Spread Eagle Ranch
and High Country Ranch, Waterton and
Bragg Creek
Tom and Carolyn Olson
403-313-6200
no website

Four Creeks Ranch, Silver Valley
Ted Buchan
780-351-2115
no website

Buffalo Horn Ranch, Olds
Peter and Judy Haase
403-556-2567
www.buffalohornranch.ca

Valta Bison Farms, Valhalla Centre
Darlene and Gil Hegel
403-237-9667
www.valtabison.com

Y is for yogourt

Vital Green Farms, Picture Butte
Joe and Caroline Mans
403-824-3072
no website

Bles-Wold Dairy and Bles-Wold Yogurt, Lacombe
Tinie Eilers and Hennie Bos
403-782-3322
www.bles-wold.com

Z is for zizania (wild rice)

Lakeland Wildrice Ltd., Athabasca
Alice and Wayne Ptolemy
780-675-4148
no website

Index

index

Acknowledgments

Just as it takes a village to raise a child, it takes a team to produce a book. My first gratitudes belong with my farmers, their families and their staff, for the many gracious generosities they have shown to me through many years of talking, eating and visiting. Hard on their heels is my first reader and toughest critic, best friend and partner, Dave Margoshes, who lights up my hours with love and joy. Love and appreciation is due to my family for raising me in an environment of knowing and valuing good food, and to my sons, Darl and Dailyn, for seeing it through to become another generation of good cooks.

The Alberta Foundation for the Arts provided grant funding that allowed me to travel through the province of Alberta, and The Banff Centre is where an early draft of certain sections of *Foodshed* was written. My sincere thanks.

The well-informed and witty Judy Schultz is the finest editor I could have wished for, one who metes out validation, honesty and wisdom in equal measure. Judy saw the value in this project from the beginning, and I bless her for her long-sightedness and for generously sharing her knowledge. My publisher, Ruth Linka, has been an invaluable support, along with the rest of the TouchWood Editions team—especially designer Pete Kohut. To Anita Stewart, the crazy-like-a-fox beacon of Canadian food, my gratitude for the years of inspiration, friendship and support. You paved the way, Anita.

Portions of this material in early versions appeared in the *Calgary Herald* and *City Palate*, *The Western Producer* and *West* magazine. My thanks to the editors at those publications.

Phyllis McCord has been my true-blue comrade for nearly thirty years, and Gail Norton has been a staunch friend and business colleague for just as long. My Slow Food colleagues likewise hold credentials as friends beyond the scope of mere friendship. To Rosemary Griebel, poet, friend, book lover, my ongoing appreciation for her friendship and finely crafted words, and to Sarah Jane Newman, thanks for the grilled cheese! More, please.

Chef, poet and food advocate, DEE HOBSBAWN-SMITH is an award-winning freelance food writer and a culinary educator "gifted with whimsy and precision." Since selling her "local fare" thirty-seven-seat Calgary restaurant in 1994, dee's writing has appeared in magazines across North America, including *Canadian Living, Western Living, Northwest Palate, West, Western Producer, Avenue, Swerve,* and Calgary's *City Palate.* Her food features and weekly column, "The Curious Cook," appeared in the *Calgary Herald* and other Canadian newspapers from 2001 to 2008. Her website, www.curiouscook.net, serves up a smorgasbord of foodie fare. *Foodshed* is her fifth book.

Dee's short fiction and poetry have appeared in *Gastronomica, The Malahat Review, The Antigonish Review, The Windsor Review, The Wascana Review, Blue Skies* and *The Society,* among others. She is working on a poetry manuscript and a collection of short fiction. After twenty-seven years in Calgary, she now lives in a one-hundred-year-old farmhouse west of Saskatoon with her partner, the poet and writer Dave Margoshes.

Dee has two sons. Both are talented and proficient professional cooks.